Urban Nature Every Day

The Four Ms
(Mum, Mat, Magic and Matilda)

First published in the United Kingdom
in 2024 by
Batsford
43 Great Ormond Street
London
WC1N 3HZ

An imprint of B. T. Batsford Holdings Limited

ISBN 9781849947527

A CIP catalogue record for this book is
available from the British Library.

10 9 8 7 6 5 4 3 2 1

Reproduction by Rival Colour Ltd, UK
Printed and bound by Toppan Leefung
Printing Ltd, China

This book can be ordered direct
from the publisher at
www.batsfordbooks.com,
or try your local bookshop.

Urban Nature Every Day

Jane McMorland Hunter
& Sally Hughes

Illustrated by Lou Baker Smith

BATSFORD

CONTENTS

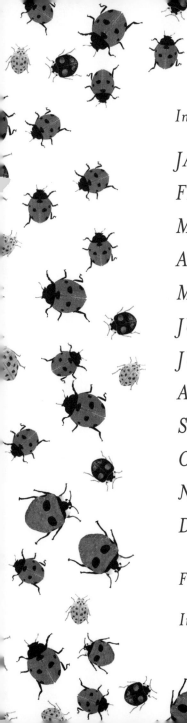

INTRODUCTION

The aim of this book is to encourage people to discover nature in urban areas. Not everyone wants, or has the time, to don wellies and waterproofs and head off across fields in search of a 'lesser-spotted something'. This book will show you that you don't need to. Nature watching (or listening) requires no specialist kit and no physical abilities and you can often visit the natural world without even stepping off the pavement. It is a common misconception that wildlife only exists in the countryside. In many areas, years of intensive farming have pushed animals, birds and wild plants out of their natural habitats and into towns and cities. Thanks to their adaptability and determination to survive, many are now thriving.

There is sometimes uncertainty as to whether a piece of land is considered urban or rural. Between the two there are wastelands of abandoned building sites, road and railway verges and roundabouts. Nature exists happily here and also deep in cities and towns: parks, communal gardens, quiet footpaths and waterways. You can visit an urban farm or spot birds from a trendy rooftop restaurant. You can watch the weather from the office window or admire miniature natural worlds in window boxes and garden containers. There are so many opportunities and what you see will depend on your chosen site. No one would expect to see a snow leopard walking down a busy road, so tailor your expectations accordingly. Best of all, go out with an open mind, ready to notice and appreciate anything you find.

This is not a field guide, but more of a starting point to awaken your curiosity and alert you to the possibilities on your doorstep. It is for beginners and experts alike. The natural world encompasses so much that it is impossible to know everything about all its myriad parts. Between us two authors, we know lots about the natural world but there is still much more to learn. We constantly surprise each other with snippets of information. Neither of us has any formal training, but we simply love the natural world and want to discover more about it. We think a pretty pink flower is pretty, whether or not you know its name. For some people, knowledge makes nature watching more fulfilling; for others, just being there is enough. Neither way is better – it's all about what you enjoy.

We have attempted to include something for every taste: large and small, good and bad, truly wild and partly cultivated. The natural world is mostly beautiful but also includes things which can lead to screams of 'Yuck! What *is* that?' You may be surprised at how interesting some things are beneath their 'revolting' exterior and indeed how those very beastly things can help make our world better.

As Brits, we are fascinated by the weather. In urban environments, the weather may often simply be an inconvenience or an excuse to change your clothes: too hot, too cold, too wet, too windy. We hope to encourage you to really look at the individual elements of the weather and marvel at them. You may then come to realize the reason for the obsession.

Connecting with nature is a great excuse for seasonal festivals and fairs: wassailing in January, cherry festivals in April and Christmas displays in a local botanical garden. At any time of year, the natural world is essential for our own well-being. The colour green is associated with feelings of tranquillity, refreshment, rest and security, and innumerable surveys have proved that we all benefit mentally from time spent with nature. There are lessons in taking joy in the moment and inspiration from nature's resilience. Never mind observing the 'lily in the field', have a look at that buddleia forcing its way through a crack high up on the wall and flowering in glorious exuberant purple profusion, attracting a bee and butterfly party.

Just as we benefit from an interest in nature so nature benefits from our increased awareness. Unfortunately, much of nature is at risk from humans in one way or another. If we are aware of plants and animals, we are more likely to want to protect them and their habitats. We hope this book will inspire you. Easy activities such as building a bug hotel or participating in a river clean-up day all help.

We both work in bookshops and talk to people all day. A customer once said, 'What I really like in cities are the trees. When they are beside the concrete, they just seem so much more wonderful. You appreciate them more than in the countryside, particularly when you think about the effort they have made to survive.' We hope this book will awaken you to the nature around you and how hard it has worked to be there.

> Let's celebrate nature in our day to day,
> At home, at work, in how we talk, how we think.
> It starts with recognising we're part of it all.
> Nature's reaching out. Let's answer the call.

From 'The Natural World' by George the Poet

Latin Names

We have deliberately not given most creatures and plants in this book their full Latin names. When watching the goings-on of the natural world, it often doesn't matter not knowing them. The bird soaring acrobatically in the sky is beautiful regardless of whether you can identify it. If you want to take your knowledge further, we have listed useful books, websites and apps in the Further Resources section on page 248.

A Warning about Foraging

There is great joy in foraging for your supper, but do be careful. Sometimes this can be dangerous, or even fatal. Don't forage from private land or eat anything unless you know exactly what it is. Think carefully about your location and what might have been there before. Railway sidings are routinely sprayed with herbicides (making berries less appetizing) and if your foraging trail is a popular dog walk, you may want to pick above the level of a cocked leg!

JANUARY

On the first day of the year, make a simple aim: to engage your
senses more and notice nature wherever you are. The trees are
bare, the weather uninviting and the nights long; once the jollity
of New Year is over, January can seem like a grim month, but
there are catkins and snowdrops to see, the hoot of an owl to hear
and the fleeting beauty of frost to admire on the coldest of days.
And rediscover wassailing, an ancient and splendid festival, which
awakens the apple trees and their spirits.

———

Little acrobat of the terraces,
we'm winged when we gaze at you

Jimmucking the breeze, somersaulting through
the white-breathed prayer of January

From 'Birmingham Roller' by Liz Berry

NATURE BY YOUR BACK DOOR

It is a myth that you need to 'go to the country' to find nature. Often, there will be amazing things right by your back door, even if you live in a seemingly barren urban landscape. A patch of rough grass by the roadside will be home to a wide variety of plants, insects and tiny creatures that will amaze you if you take the trouble to look at them properly. Admittedly, you may not spot a tiger or even anything particularly memorable, but it will show you a world both different and wondrous, which exists alongside the one you inhabit.

2 JANUARY

FROST

Frost forms overnight when temperatures fall and the moisture in the air freezes. By morning everything is coated with a thin layer of ice crystals. The air in these crystals makes the frost appear white. If the weather is very cold, fern frost patterns form on windows, creating beautiful and delicate swirls. Frost rarely lasts long, particularly in urban areas, as it is only a very thin layer and will easily melt. If you want to see your local park transformed into a genuine winter wonderland, you will need to get up early but – trust us – it is well worth the effort.

3 JANUARY
WATCH THE YEAR

All naturalists have 'their patch'. It probably won't be exclusive to them, but it will be somewhere that they know well, well enough to notice subtle changes as the seasons progress. It needn't be anywhere spectacular, just a place that has things that interest you. Visit your patch as often as you can: every day is obviously ideal but, for many people, this may be totally impractical. Once a month is fine, once a week is better. The rate at which things grow will adapt to the light, water and warmth; the activities of the birds will alter as they mate, raise a family and possibly prepare to migrate; and the animal and insect visitors will vary. Don't be put off by the weather, some of our best sightings have been while huddled beneath a brolly.

4 JANUARY
ALDER CATKINS

Alder trees need light and prefer wet 'feet', so they can be most often found along rivers and canals. Each tree bears male and female catkins: males are drooping and purple, gradually turning to yellow in spring; and the females are like tiny fir cones at the tips of the branches, turning from bright green to dark brown as they ripen. You won't be able to see the most important thing about alders; they use the carbohydrates they produce during photosynthesis in an incredibly clever way that improves the quality of both the air and soil around them. The other unusual feature is that the wood does not rot as long as it is kept wet – which is why alder posts are, almost single-branchedly, holding up Venice.

BUG HOTELS

Bug hotels may look like stylish garden art, but they are actually a practical way to help bees and other insects. There are many species of bee who live alone and do not produce honey. These solitary bees build and provision their own nests, rather than living in a community hive. For them, and insects such as woodlice or lacewings, cities are often hard, inhospitable spaces where they can struggle to find winter shelter. If you have a city flat with a tiny balcony, think about putting in a bug hotel. As with all hotels, location is key and a sunny spot away from frosts is essential. Use a hanging basket, recycled pallet or a drainpipe and fill with a mix of found objects: drilled logs, canes, pinecones, straw. Make it an attractive pattern to please yourself, but make sure there are plenty of nooks and crannies to please the insects.

FOX

When April comes around there will be cute fox cubs at play. At this time of year, the fox world is also with us, but is much less appealing. How often have you been woken in the night by blood-curdling screams reminiscent of the soundtrack of the more lurid crime dramas or horror films? The chances are, unless you are unlucky enough to live in a particularly bad part of town, this will be foxes. Foxes scream and bark to communicate with each other, and at mating time in January this is at its peak. Males scream to warn off other foxes and protect their territory, while females scream to attract a mate. This activity tends to occur at night and neither tod nor vixen care that you have an early meeting.

RECORDING NATURE

Keep a nature diary – start one, today. This will be an easy resolution to keep as the more you write the more addictive it will become. It doesn't need to be a fancy diary and you don't even need to write it every day, although the habit helps, but simply jot down anything interesting you see. Over the years you will see bluebells arriving earlier or later, migration patterns changing and heatwaves repeating themselves. Photos can record the moment but writing in a diary makes you think more than the mere click of a button. You will soon find unexpected and fascinating connections across the natural world, in your own back garden, park or far beyond.

GREAT TIT

A really pretty great tit is a regular visitor to our bird feeder. It's great in tit terms, but these are small birds. Tits (great, blue, coal and all the tits in between) are always welcome. About the size of a robin and with a similar personality, tits are comfortable with being watched and flit about catching insects on the wing. Great tits have yellow breasts with a black stripe down the centre, greenish backs and black heads with very white cheeks. Blue tits are the ones with the little blue caps, which they can raise as a crest if they are in the mood, while coal tits look similar but have greyish chests. All of them are pretty vocal and like to chat among themselves.

IT'S NOT THAT EASY

Nature programmes, podcasts and videos are great, but they can give a false impression of how easy it is to find wildlife. The camera operator may have been sitting in a particular spot for hours, but we only see the three minutes of perfectly edited film. Equally, none of the experts are ever seen peering at something saying, 'I'm not sure. It could be ...' None of this helps when you go to a carefully researched site, certain of a glimpse of a particular creature, only to find it refuses to appear. Don't lose heart. Watching nature does not come with a guarantee that you will always find what you are looking for. Look around, notice everything else and simply appreciate being outside in the wild.

SITTING STILL

For much of your time spent nature watching you will need to learn to sit still. Very still. Many of the creatures that would normally run away may notice you, but as long as you are not moving, you do not pose a threat. Others, which would run away anyway, may not notice you or, at the very least, sitting still will earn you some time before they do spot you. Find somewhere comfortable, settle down and watch the world around you. Try to *look* and *notice* rather than allowing your mind to wander or even close down; this is not a mindfulness or meditation session. Don't set a time, just watch the natural world.

A tip we learned from Simon Barnes: always take a plastic bag to sit on, it's hard to be still with a wet bottom.

SNOWDROP

Strictly speaking, most snowdrops you see in urban areas will be garden flowers rather than truly wild, but they are the earliest of the spring bulbs, so we are not going to let a technicality stop us from including them. The pretty white flowers hang like tiny lanterns from thin, green stems and are one of the markers that winter is beginning to lose its grip. Measuring 7–15 centimetres (2¾–6 inches) high, each plant has three to four grey-green, grass-like leaves. The three inner tepals (where petals and sepals are indistinguishable) have green tips, the outer three are white. Single flowers, looking like miniature shrouds, can represent death, and according to ancient lore should not be brought indoors, but bunches are a sign of purity and virginity.

Larger, later to bloom and with six green spots are snowflakes, also garden flowers and equally pretty.

FEED THE BIRDS

Channelling your inner Mary Poppins supports birdlife and encourages them to visit. While feeding in winter is obvious, food shortages can occur year-round, so provide a regular supply.

Bird tables and hanging feeders are excellent. You are providing a treat, not setting bait, and ground-level feeding exposes birds to predators. While birds mostly eat on the spot, they sometimes take titbits back to the nest, so think about what you are putting out. Young birds can choke on larger morsels, so avoid whole peanuts and large pieces of bread or fat, and look carefully at seed mixes – some contain split peas, lentils or uncooked rice all of which are only suitable for larger birds.

The ultimate treat of the bird world is the deliciously named Flutter Butter, a low-salt avian peanut-butter popular with tits.

KIT YOU WILL NEED

On a basic level, all you need to notice nature are your own senses, but if you want to record what you see you'll need a little kit. A note pad, pen and pencil or a phone are perfect to document what you see. There is no 'best' way to record your finds; it is entirely up to you. You can identify what you see using a field guide or an app, or both if you prefer. You can take photographs or draw. What you won't need are specimen jars; as a general rule, look but don't disrupt, is the best guideline.

RED SKY AT NIGHT

Red sky at night, shepherd's delight,
Red sky in the morning, shepherd's warning.

Whether you say shepherd or sailor matters little: if you live in the British Isles this is an accurate weather forecast, and is probably more accurate than many of the countless websites and apps we rely on. A red sky at sunset means high pressure is coming in from the west, bringing dry weather. A red sky in the morning means the high pressure has moved east and is likely to be followed by wet low pressure. Of course, this does not take into account the gardeners and farmers who would delight in rain during certain times of year, but then the saying has never, as far as we know, catered for anyone other than shepherds and sailors.

15 JANUARY
START SMALL

If you are new to watching wildlife or even if you need a new inspiration – start small. We don't mean only look at small things, although you can if you wish, but rather find a single thing you are passionate about and concentrate on that. One thing may lead to another, and your interests may widen, or you may remain focused on a particular aspect of the natural world – birds, bees, butterflies, trees, flowers or grasses, it really doesn't matter. Remember you are doing this for fun, not for an exam, and also remember that, however niche your interest, nature will surprise, delight and possibly even shock you.

OWL

We fell in love with owls when reading children's stories: *Winnie-the-Pooh*, *Harry Potter* and the superlative *The Owl Service*. Traditionally a wise old country bird that lived in the woods, owls are increasingly moving into town, particularly during the winter when food is scarcer. The best places for an owl prowl are parks and cemeteries, but anywhere with well-established large trees or even telegraph poles offers the chance of spotting a roosting owl. The open spaces of sports grounds and golf courses also attract owls, as it is easy for them to spot their prey scurrying across the grass.

The owl's distinctive round, flat 'face' is surprisingly to do with its hearing. The slightly concave discs on owl's faces collect sound and direct it towards their ears. Try it out yourself – if you cup your hands behind your ears, you will mimic the concave planes of an owl's face and actually hear better. Hoot if you agree.

WASSAILING

Everyone needs a bit of jollity in January and that is exactly what wassailing provides. The oldest or best tree in the orchard is chosen as Apple Tree Man, or the guardian of the orchard. Cider is poured on the roots and a piece of toast or cake, soaked in cider, is placed in the branches to attract robins, which are the good spirits of the trees. Guns are fired or saucepans banged to scare away any bad spirits and wake the trees, which are then often serenaded with traditional songs. Finally, there are toasts (with more cider) and a good harvest is ensured. The name is thought to come from the Anglo-Saxon *wes hal, was haile* or *wase hail* meaning to be in good health. Any urban apple tree would welcome such attention.

The festival usually takes place on 17 January, which was the date of Twelfth Night before the introduction of the Gregorian calendar in 1752.

OBSERVATORIES

Normally, we advocate getting up close and personal with nature but sometimes a little help is required. While in theory observatories are sited on hills away from towns to avoid light pollution, there are a surprising number of them in or near to cities, and many have tours or public nights. Visit gostargazing.co.uk to find one near you. If you get the chance, do try to visit and gain a cosmic perspective on the night sky. As well as powerful telescopes enabling you to see the surface of planets or more distant stars, observatories have helpful and knowledgeable staff who can explain just what you are looking at.

GALES

There is something so exuberant about a good gale. One which, as they say, blows the cobwebs away. The word gale derives from the old Norse *galinn*, meaning mad, frantic or bewitched, and these strong winds are powerful and even violent. If you are not one to run out into the gale and embrace the feeling of being buffeted around as the air turns your cheeks pink, then enjoy the fun from inside. Peer through your windows and look at the trees, the cranes and any cabling as it dances in the wind. Or listen to the Shipping Forecast on the radio – 'Warning of gales in Stornoway' sounds exciting and romantic (unless you happen to be in Stornoway).

LEARN TO LISTEN

Living in cities we are surrounded by noise. It is almost as if we fear silence, so we fill the void, but in order to cope with the continual bombardment we have learned to block out much of the constant din: traffic, other people, music in shops. Go to a park or garden, sit (or lie) with your eyes closed and listen. You will hear cars and people, but go past those and listen to the wind in the trees. Birdsong is beautiful and uplifting, even if you don't know exactly which bird is saying what. Hearing something is every bit as important as seeing it, and in some ways even more so as you can allow your imagination to run riot at the same time, which is always a good thing.

21 JANUARY
JELLY EAR

The elder tree in the local cemetery has a nice crop of jelly ear
fungus lined up like pellucid brown ears listening out for rain. Jelly
ear or wood ear is one of those fungi that is fun to spot because of
its unusual appearance. From a distance it looks like a party frill
on the trunk of a dead or dying tree, but closer in it really does
look disconcertingly like a cluster of ears, the colour and texture of
rubber bands. One of its other common names is Judas' ear, possibly
due to its frequent appearance on elder trees, reputedly the tree from
which Judas hanged himself after betraying Jesus. In 17th-century
herbal medicine, jelly ear was used to treat eye and throat conditions.

22 JANUARY
WINDOW BOXES IN WINTER

It is easy to have spectacular containers in spring, summer and even
autumn, but winter requires a little more ingenuity. Walking around
any town at this time of year will reveal a high number of sad-looking
containers, either empty or sporting the collapsed remains of last
summer's flowers. Birds, insects and even humans need flowers
and greenery now more than ever. Evergreen grasses, heathers,
wallflowers and open flowers such as hellebores and pansies are good
choices. Seed heads are also good but only as long as they remain
stately and upright. Bulbs can be planted underneath all these and
will push their way easily through the greenery in spring.

NOTICING NATURE

Some people notice more than others, but it is very easy to train yourself to be more observant. There are two ways of looking at something: a sweeping view that gives a good overall impression of the area; or a slow, detailed look that takes in all the little things you might otherwise miss. Both are equally important. You also have peripheral vision, which you probably use without thinking about it: checking for traffic while making sure the bus you want to catch isn't pulling away; keeping an eye on the frying pan while chopping the onions. Learn to use it when you are out and about; look at the tree you are trying to identify but also be aware of the bird flying out of the branches.

NO SUCH THING AS BAD WEATHER

Alfred Wainwright wrote a number of walking guides to the Lake District, but for our purposes his most important piece of wisdom was, 'There's no such thing as bad weather, only unsuitable clothing'. Those who have had picnics ruined by an unexpected storm may well disagree but, if you have the right clothing – or a decent umbrella – even soggy picnics can be unexpectedly successful. Invest in a decent waterproof jacket and comfortable waterproof boots, walking or wellies as you prefer, in cities it doesn't really matter. We prefer standard umbrellas to the huge golfing ones, which are unwieldy and antisocial. A good brolly will allow you to make notes, sketch and peer at something in detail and, if the rain continues, you can eat your lunch or tea beneath it.

BINOCULARS

Nothing makes you look more like a serious naturalist than a pair of binoculars – or 'bins' as you will learn to call them. But, joking apart, binoculars are truly amazing, whether you are interested in stars, birds, tree-tops or even architectural details on tall buildings. You can also get close-focusing binoculars for looking at flowers or butterflies in detail. Like many things you get what you pay for, but a reasonable pair needn't cost a fortune. Go to a reputable shop, ask for advice and test as many pairs as you can. Once you have a pair of bins, *always* take them with you; they are no use sitting at home.

26 JANUARY
HAZEL CATKINS

The drooping yellow catkins appearing on hazel trees now are male, and are the distinctive markers to use for staking out trees from which to harvest nuts later in the year. The edible nuts, also known as cobs or filberts (filberts are long rather than round), which ripen to a dark brown and are ready to harvest in autumn, grow from the not-particularly-noticeable female flowers. Hazels tend to be shrubby rather than majestic, but manage to be categorized as trees as they sometimes grow on a single stem (the rough divide is shrubs: many stemmed; trees: single). The purple filbert (*Corylus maxima* 'Purpurea') is a popular urban tree with pretty, purple-tinged leaves and catkins.

The green nuts are soft and should be eaten when picked; the mature brown ones will keep for several months.

27 JANUARY
CAPTURING THE MOMENT

The next step from taking time to appreciate nature is to try and capture the moment. Painting, photography and writing are all effective ways to take 'just looking' to another level and for a keen artist the natural world is a perfect subject. With photography there can be a tendency to take a quick picture with your phone and never go back. Resist this. Examine your subject: think about framing. Close up or panoramic? Are filters and effects appropriate? How would it look in black and white? If you draw or paint; what medium is best suited to your subject? Use your art as both a means to create something beautiful or arresting and as an activity to deepen your relationship with the subject.

BIG GARDEN BIRDWATCH

Since the Big Garden Birdwatch began in 1979, some 139 million birds have been counted. It is the largest wildlife survey in the world, and taking part is both fun and the opportunity to play a part in an important conservation activity. It only takes an hour. Pick an hour on Birdwatch weekend – the last weekend in January – and make a note of the birds you see. It doesn't matter where you are – you can watch from a balcony, garden or in a park. Remember to only count the birds that land and report your results on rspb.org.uk/get-involved/activities/birdwatch. If you need a little help to identify the birds, then register in advance with the RSPB and they will send you a free identification guide. The data collected helps build up a picture of how different species are faring, and which ones may need help.

RAT

Cycling down the towpath there is a flash of brown as something darts in front of you – it might be a rat. It is one of the urban myths that in Britain 'you are never more than 6 feet away from a rat'. In truth, it is more like 164 feet, but even this is averaged across the country, and rats are not evenly spread out – there are more in sewers, on farms or in grain stores, and the chance of you meeting one at home are slim. They will be having fun elsewhere. Rats are really very social animals who share food and care for each other. A group of rats is called a 'mischief', a name which reflects their fun-loving side.

30 JANUARY

JOHN LEWIS-STEMPEL

Military historian and farmer John Lewis-Stempel is one of the heroes of the literary natural world. *The Times* described him as 'Britain's finest living nature writer' and we agree. Many of his books relate to his traditional farm, far from any city, but he has also written delightful little books on owls, oak trees and foxes. *Nightwalking* is one of our favourites; the seasonal walks take place in west Herefordshire but will encourage you to regard the dark with a new sense of wonder and perhaps even take a detour away from well-lit areas on your way home from the shops, office or pub. There is much to see in the dark.

31 JANUARY

RAVEN

From the Bible to Edgar Allen Poe, ravens occur frequently in folklore and literature. We read about the princess with hair as black as a raven's wing; we fear that if the ravens leave the Tower of London the kingdom will fall; we understand ravens are the messengers of the gods and that the raven was the first bird Noah released from the Ark. One of the most intelligent of birds, ravens are somewhat unfairly portrayed as symbols of ill omen, despite their playful nature. Being scavengers by nature, ravens were originally attracted to cities by the rich pickings around meat markets and slaughterhouses but made themselves very much at home, and today are more likely to be found scouring rubbish bins for that leftover piece of burger or kebab.

FEBRUARY

February is the month we start to get an inkling that winter may be ending, as everything begins to wake up. Notice the fat buds on the trees just waiting for their moment or examine carpets of rotted leaves with green shoots just pushing through. By the end of the month, crocuses will have joined the snowdrops and cyclamen as the first of the spring flowers. There is more birdsong to listen to and, on warmer days, you may even catch sight of a bee. Before all the leaves come out, take time to admire the shape of trees looming eerily through the fog, or the lacy lichen creeping along branches, and any cold clear nights are great for stargazing. Warm up your nature-spotting muscles, as there will be more to come as the year progresses.

———————

Magically awakened to a strange, brown night
The streets lie cold. A hush of heavy gloom
Dulls the noise of the wheels to a murmur dead:

From 'Fog' by Laurence Binyon

1 FEBRUARY
CROWDED TOGETHER

Everyone knows that human population densities are greater in
urban areas than in the country, but what may surprise you is that
wildlife densities are also greater. One reason is thought to be
that many animals and birds have fewer predators in cities or are
better able to escape them. Creatures that adapt to city life may also
be better at surviving than their rural counterparts; their ability
to thrive on a wider range of foods means that they need smaller
territories. This often makes it is easier to spot wildlife in cities:
in the country you only know a fox has visited by the chaos it leaves
behind; in a city you are quite likely to meet one on your way home
from a night out.

2 FEBRUARY
URBAN FOREST

Forests can conjure up images of huge wild areas inhabited by bears,
wild boar and the like, but everything from a stately horse chestnut
down to a single blade of grass is part of an urban forest. This
type of forest forms the network of greenery within a city; street
trees may stand as individuals, but they create pathways for wildlife
through built-up areas. Trees in private gardens, parks and along
waterways add to this chain, which also helps to control stormwater
and air pollution. This is a forest you may not notice until you look
(and then we hope you will be pleasantly surprised at the amount
of greenery surrounding you), but it is one that makes cities more
attractive, healthier and altogether better places for everyone.

FEATHERS

Now we no longer have dinosaurs, only birds have
feathers – but what a wide variety of colours, sizes
and shapes of feathers there are: from humble
little curls of white down to exotic peacock feathers.
While feathers, like human hair, are formed from
keratin, their structure is more complex, and they
can be controlled by tiny muscles in the follicles
in the bird's skin. Their function is to aid flight,
to protect the bird from bruises and scrapes,
insulate from water or cold, provide camouflage
or help a bird stand out in an arresting mating
display. Birds have a lot of feathers. The
tiniest songbirds can have between 1,500
and 3,000 feathers, while an adult
swan may wear up to 25,000.

LICHEN

Lichens are a combination of an algae and a fungus. They grow on
trees, rocks and stonework, appearing as patches of grey, brown,
mustard yellow or white. Sometimes they resemble moss, sometimes
the very stones they are on. Lichens are hard to date exactly, but
many are older than a hundred years, the oldest on record starting
its life over 8,000 years ago. They can cope with frost or drought,
but dislike pollution; it is heartening that they can now be found in
many cities.

ENTHUSIASM

Watching the natural world often requires one to be still and quiet for long periods of time. Plants won't (or can't) run away if one makes a noise, but an inappropriate yell along the lines of, 'Look! They're over here!' will cause wildlife to scarper as surely as if an eagle has landed nearby. So, an element of caution is needed, but that doesn't mean nature watching has to be serious – it's fun, enjoy it and, most importantly, celebrate your finds. A rare sighting is great; you may think you would prefer to keep it for yourself, but actually a sighting shared is much more wonderful. Talk to other naturalists – they will all want to tell you what they have (or have not) seen and you may well learn something of interest.

BARK

Bark is often the easiest way to identify a tree, particularly at this time of year when many trees lack leaves, flowers or berries. It is amazing stuff; it provides a hard outer skin that protects the tree while expanding at the same time to allow the tree to grow. Remove too much and the tree is likely to die. Planes and silver birches shed patches of bark to rid themselves of pollutants; cherries have obvious horizontal rings; the oak's bark is rough; while that of sweet chestnuts spirals round the trunk with deep fissures. Nearly all trees have lenticels, most of which look like tiny breaks in the surface of the bark; these allow the tree to take in extra oxygen above that which they make from photosynthesis.

LOOK AT WHAT IS THERE

Going out to look for something specific and not seeing it can be disheartening. Someone may tell you they have seen a badger on the local golf course, so you set aside an evening and spend it hanging round the greens staking out what looks very like a sett. Nothing. You are told about a wonderful crop of blackberries only to get there to find someone has beaten you to it and taken the lot. Do not lose heart. Instead look at what *is* there. There may be other birds around the golf course, or squirrels larking about near the hedgerows. Notice the silhouettes of the trees as the sun sets. Put your foraging basket away and look properly at the bramble plant, perhaps even draw the flowers. Be an opportunist and appreciate whatever nature chooses to show you.

GADWALL

You are most likely to see urban ducks on ponds in parks rather than truly living wild, but that doesn't make them any less interesting. The gadwall is smaller than a mallard and both the male and female have patches or a speculum of white on the undersides of their wings, which is most easily seen when they are flying. The males are grey-brown with distinctive black tails when they upturn themselves to dabble, while the females look similar to female mallards. They are smaller, but the easiest way to be sure is if you see them alongside a male gadwall. Only a few gadwalls live in Britain, but a great many spend the winter here, so that is the best time to see them.

PLANETS

Even in the city, planets are visible if you look carefully. If you like
to know what you are doing, then one of the best websites around
is theskylive.com. Once you have put in your location, the site tells
you what you should be able to see that night, when to look and
whether you can see it with the naked eye or need bins or a telescope.
There are even maps. However, you can still take enormous pleasure
from the beauty of the night sky even if you cannot tell your Jupiter
from your Mars (or either from a night-flying plane). Just gaze
heavenward and romanticize about what may be up there and if you
will ever visit in your lifetime.

EVERYTHING THAT WILDLIFE NEEDS IS ON THE DOORSTEP

Cities make surprisingly good habitats for a lot of wildlife: they
offer water, food, shelter and often an escape from the uncongenial
agricultural world of large, weed-free fields surrounded by open
fences rather than hedges. Of course, not all the countryside is like
that, and much urban land is distinctly hostile, but a great many
birds, animals, plants and insects survive and even thrive in built-up
areas. An important factor, which can be either good or bad, is that
urban areas tend to be warmer than the surrounding countryside as
buildings absorb heat by day and release it at night. This gives many
plants a longer growing season with a knock-on effect for the insects,
birds and animals that depend on them.

HOUSE SPIDER

House spiders are the ones you find in corners, under the
sofa or behind the fireplace: wherever you haven't swept
recently. If you find one in the bath, the chances are it has
fallen in, they don't emerge from plugholes. The hairs on
their legs, which help them climb walls, just can't get a grip on
the smooth enamel, so be a sport and help the little fellow out.

House spiders spin funnel-like webs where they wait
patiently for small insects, or as they might prefer to say 'lunch',
to arrive. They don't go out much, but when they do, they motor.
A house spider can cover 0.5 metre (1½ feet) in a second and is
one of our fastest invertebrates. They are brown, sometimes with a
herringbone pattern on their body, and quite large but completely
harmless unless you happen to be a fly.

IT DOESN'T NEED TO BE RARE

Many plants and creatures are called common, but that doesn't mean
they are any less interesting or beautiful. Now, with so much of our
open spaces under threat, it doesn't even necessarily mean they are
very common. Don't ignore something just because it is easy to see;
daisies pop up in every patch of grass, robins sing in every park or
garden, and spiders' webs festoon plants in autumn, but that doesn't
make them any less remarkable. Rejoice in spotting the rare and
unusual, but don't ignore the commonplace.

SILVER BIRCH

The silver birch is one species of tree that is instantly
recognizable by its bark. Young stems start brown, but
their bark turns white as they grow into branches, and
the trunks of the trees develop dark fissures in the papery
white surface giving a rough feel. The leaves are triangular
with double-toothed edges and the trees bear male and
female catkins; the male growing like yellow lambs' tails in
autumn and the female, shorter and bright green, following
in spring. These thicken once pollinated, turning dark red
and releasing hundreds of tiny seeds in autumn. In Celtic
mythology, the tree represents renewal and purification and
a birch besom or broom is believed to purify your garden as
well as tidy it.

A POSY OF WINTER FLOWERS

Nothing says 'I love you' quite as much as a little posy of flowers
on the breakfast tray. Well, obviously champagne, chocolates
and diamonds might also do the job, but they have less
relevance here. This might seem a thin time of year for flowers,
apart from the hot-house forced specimens found in florists
and supermarkets, but there is a surprising variety available if
you look. Many flowers at this time of year are small: bulbs such
as miniature narcissi, reticulata irises and crocuses, or florets
of early blossom. The branches of shrubs such as witch hazel,
Christmas box and rosemary can provide a twiggy structure to
support the flowers. Use a narrow glass or small vase and you
will only need a few flowers to give sparks of brilliance.

GOLF COURSES

Like urban parks, golf courses are welcome oases for wildlife, in fact recent studies have suggested that they support greater biodiversity than farmland. Greenkeepers keep them well-watered in all but the most extreme droughts, so early and even late-to-the-party, birds can get those tasty worms. The rough, while a challenge for the golfer, is a great spot for wild grasses, butterflies, beetles and other insects. Foxes and even badgers roam at night, safe from traffic. Courses are often surrounded by trees, no doubt to prevent enthusiastic but less able golfers from hitting windows of neighbouring houses, but they provide a haven for birdlife. If you play yourself, or have a public footpath through a course, keep an eye out, you will be surprised at what you may see.

LETTERS AND DIARIES

Reading nature letters and diaries is informative, inspirational and the perfect activity for a wet or sunny afternoon by the fire or in a deckchair respectively. Our favourites are by Gilbert White (1720–1793) and Francis Kilvert (1840–1779). Both vicars, they lived in rural England rather than cities, but we include both writers here for their infectious delight in the natural world. Unusually for the time, Gilbert White believed in studying birds and animals in their natural habitat, so his letters paint a vivid picture. Francis Kilvert's diaries cover everything from how to press flowers to dancing, to the delights and hardships of the various seasons. They both describe a world long vanished but one we would happily visit.

17 FEBRUARY
TOAD

Despite their brown warty appearance, you should welcome toads to your garden, as they feed on slugs, snails and other enemies of the proud gardener. They breed in ponds, but most of the year will make a home in gardens: in flower beds, under flowerpots, in the woodpile. You can sometimes see mass toad migrations in the early spring when, on warm, damp evenings, gangs of toads (technically the collective noun for toads is a 'knot') travel determinedly back to their breeding ponds. There are even groups of volunteers who monitor known 'toad crossings' on busy roads in the breeding season to ensure that the single-minded migrating toads are not mown down by cars. Toads don't hop (that is frogs), they walk.

18 FEBRUARY
PLANE TREE

A hybrid of the oriental plane and the occidental plane, the London plane can be found in most cities. It was the first tree to be planted systematically in London in the mid-19th century, giving rise to its common name. It survives the harmful effects of pollution by shedding patches of bark and has smooth, shiny leaves that are washed clean by the rain. The leaves are large, palmate and lobed and the flowers spherical: the female is reddish-brown, the male pale green. The ripe fruits are brown and remain on the tree throughout winter.

EARWIG

Despite their name you are unlikely to find these little creatures in your sister's ear. They prefer dark cracks and crevices, and are often found around window boxes or planters or in dark corners of the garage, mainly coming out at night or when a pot is accidentally kicked over. Looking like a very slender, glossy brown beetle with rather fearsome horns, they do have wings but rarely fly, preferring to scuttle about. Earwigs are mainly vegetarian, feeding mostly on decaying plant matter, although they will occasionally munch on aphids. Unlike most insects, female earwigs are devoted mothers that care tenderly for eggs and larvae.

OAK GALL

Over 50 species of gall can be found on an oak tree, ranging from hard, circular growths on branches, to tiny specks on the underside of leaves. They are one of nature's mysteries, with cecidologists studying nothing but galls and coming up with not many answers. The eggs of tiny parasitic wasps somehow affect the growth of the tree so it forms a protective coat for them. Oak apple galls have the most prestigious history. Until late in the last century, in some countries, they were ground and mixed with chemicals to produce ink for almost all-important documents, including the Magna Carta and the American Declaration of Independence.

21 FEBRUARY

THE POETRY OF NATURE

A great many poets have written about the countryside and nature, with John Clare and Gerard Manley Hopkins being two of the best known. However, many city-dwelling or visiting poets have written just as lyrically on urban nature including Amy Levy, Andrew Motion, Alfred Noyes and Laurence Binyon. Even John Clare visited London's Epping Forest. Reading and nature watching are not necessarily compatible, particularly if you become engrossed in your book to the exclusion of all else around you. Reading poetry is another matter; the nature of most verse is that you read and then stop to think, at which point you look around. Poetry can make you notice more, not less, of the natural world.

22 FEBRUARY

SHREW

Shrews are like tiny little mice with long, pointed snouts and bright, beady eyes. They hang out in the long grass in gardens or on commons, but as they are so small and quick they can be difficult to spot. They are noisy little critters though, and if you listen carefully, you may be able to hear their high-pitched and almost incessant squeaking. Despite scoring highly on the cute scale they have very sharp pointed teeth that they are not afraid to use. They are constantly ravenous and scurry about on an endless quest for food pausing only briefly to nap. Shrews need to eat every 2–3 hours or they will die (some days we just know how they feel).

PIKE

The pike has a fearsome reputation as an aggressive predator. It is the largest freshwater fish in the UK, living up to twenty-five years and growing up to a metre in length with that formidable set of sharp teeth. They favour deep weedy slow-moving water, but are common in canals and round docks where they lurk under piers or moored boats from whence they burst out to take their prey. They feed on smaller fish, frogs, small mammals and birds. Young pike are called jack (all of them, no Jills).

SHOVELER

Shovelers are easily recognized by their flat, shovel-like bills, which they use to filter tiny creatures and seeds from the water or mud. Unlike most other surface-feeding ducks they rarely upend, instead dabbling along with their bill just on the surface of the water. They like shallow water with plenty of plants for both cover and food. The male has a dark green head with a black, white and chestnut body, while the female is brown. Both have panels of pale blue on their upper wings. On land, they look cartoon-like with clumsy movements and their huge bills seeming to unbalance them, but they are graceful in the air and perform courtship flights round their territory in the spring.

25 FEBRUARY

PIED WAGTAIL

There is a chittering from the supermarket car park and sure enough a flock of pied wagtails has roosted there this evening. There is something about supermarket car parks and shopping centres that seems to attract wagtails – possibly the warmth and the lights or maybe they just like shopping? They are really perky little birds with their jerky, hopping walk and leap-frogging flight and, of course, the long wagging tail that gives them their name. There are three types of wagtail in Britain – pied, grey and yellow – but in town it is the black-and-white pied wagtail that is most common. They build their cup-shaped nests from grasses and moss in the oddest of places: in cracks in walls, behind ivy or even up drainpipes or in scaffolding.

26 FEBRUARY
CROCUS

Crocuses are primarily garden plants, but they earn their place here
for being one of the markers that nature is waking up. And, given
half a chance, they will naturalize outside the garden gate. Lilac and
white or egg-yolk yellow are the most common colours, the yellow
flowers being slightly larger and later. Planted or wild, crocus carpets
will make you stop in your tracks. A cousin, the autumn crocus, will
create a similar display later in the year.

27 FEBRUARY
GREEN BELT

The urban environment in Britain is blessed with parks and green
spaces, and as we have shown there is plenty of opportunity to
interact with nature, even in the most built-up areas. There are also
tracts of land designated as green belts, which act as a buffer between
town and countryside. The concept of a green belt was introduced to
reduce urban sprawl, provide recreational spaces and assist in nature
conservation. The Metropolitan green belt around London was first
proposed in 1935; since then the practice has expanded around the
country and there are now 14 official green belts. If there is one near
you, do try to visit.

28 FEBRUARY
EARTHWORM

Worm-rich soil is healthy soil, so we always like to see a little wriggler when digging in the garden. Earthworms have segmented bodies with tiny, fine bristles that help them to move themselves along their burrows. The 'mouth' of the worm takes in the soil, passing it through their bodies and extracting nutrients from any plant matter, so for worms digging and feeding are simultaneous. If not actually digging yourself, you are most likely to encounter worms when it rains, as that is when they come up to ground level. Worms need to breathe through their skin and if their burrows fill with water they can drown, so at the hint of a downpour they head up to the surface.

29 FEBRUARY
IT'S A HARSH WORLD

'Nature, red in tooth and claw' is a quote from the poem *In Memoriam A. H. H.* by Alfred, Lord Tennyson, but it is an important reminder that not all nature is fluffy lambs and pretty flowers. Each species has its own rules for survival and some of these may seem harsh: baby birds pushed out of their nests, animals eating other animals and even plants having the goodness sucked out of them by predators. Wildlife in urban areas lives close beside us and we may see things that we don't like. Should we intervene? On the whole, we think not advisable in most cases as all nature depends on a balance, one which humans can all too easily upset. You may save the little mouse, but what about the owlet waiting in the nest, hungry for supper?

MARCH

March is the month of the spring equinox, the moment in the calendar when our days start to be dominated by light rather than dark. The changing of the clocks gives us even more time in the evening to get out (focus on that and you'll mind less about losing an hour in bed). The cold, possible snow and bracing winds mean nothing to the daffodils and, towards the end of the month, hedgehogs and adders will wake up and birds will start their ritual mating and nest-building activities.

———————

A cold wind stirs the blackthorn
To burgeon and to blow,
Besprinkling half-green hedges
With flakes and sprays of snow.

From 'Endure Hardness' by Christina Rossetti

1 MARCH
DAFFODIL

There are the institutional daffs, planted in neat groups of single varieties and there are the tearaways, mostly planted rather than truly wild, but popping up in unexpected places or mingled together in a glorious range of heights and hues. Everyone (we hope) can recognize a daffodil, but look closely: there is every shade and colour combination, from the purest white through the palest lemon yellow to deep orange. There are trumpets heralding the arrival of spring and flat flowers holding their faces up to the sun. All daffodils seem to be born with in-built raincoats; while other, more delicate flowers hide away at this time of year the daffs are out there smiling, whatever the weather.

2 MARCH
PEREGRINE FALCON

Faster than a speeding bullet, able to nest in high buildings, Superman, oops... the peregrine falcon, is making itself at home in British cities. They have taken to cathedrals and bridges, and even Tate Britain and the Houses of Parliament in the heart of London. For the peregrine falcon, tall buildings are just oddly symmetrical cliffs and the abundant pigeons found in cities provide a plentiful supply of food. Peregrines nest on ledges and lay eggs in March and April from which hatch the most delightful, fluffy chicks, but you are unlikely to see a nest close up, as nests are protected and must not be approached. However, the urban bird spotter will be able to watch the adult falcons swooping through the sky at an impressive 320 kilometres (200 miles) per hour.

3 MARCH
BARK RUBBING

Bark can be very beautiful, but you should never pull it off a tree; it is the all-important protective layer. Instead, make a bark rubbing – something you may have done as a child and not thought about since. Obviously, first stake out your tree. Then choose a dry day. Place a sheet of strong, thick paper over a section of interesting bark, hold it in place with one hand and rub over gently with a stick of wax crayon or charcoal. A4 or smaller is easiest if you are on your own, but with multiple hands available to hold the paper in place you can create wrapping paper or, probably, wallpaper. We just haven't tried that yet.

4 MARCH
THE BLACKTHORN WINTER

Blackthorn trees are rounded, reaching up to 4 metres (13 feet), but are usually smaller and look like large, scruffy shrubs. Their dark, almost black stems are dense, twiggy and thorny and provide a perfect backdrop for the beautiful white blossom which appears along their stems in late winter or early spring. This time is often called the blackthorn winter, a period of bitter cold after a false spring or unseasonably warm spell. It probably shocks the blackthorn as much as us, although blackthorn blossom minds frost less than other fruit trees. Notice the trees now and mark their positions; they become inconspicuous during summer, but in autumn they bear the gin-infusing berries known as sloes, which are well worth harvesting.

COOT AND MOORHEN

When we feed the ducks, we play 'Is it a coot or a moorhen'? Both are keen to grab their share of any duck food on offer. To win, you need to know that the coot is the one with the white bill and the white shield on its face, while the moorhen has an orange bill with a yellow tip. But be careful, because the shield on the face of some coots can also be a red or yellowy colour – look at the bill! Coots are a little larger and rounder too, although size is not the best identifier as there is some overlap in subspecies. They both have very big feet which, despite them being excellent swimmers, are not webbed.

FROGSPAWN

Rafts of frogspawn hang in the shallows of the pond in the park. If you approach and look closely, you can see each little gelatinous egg contains a tiny, black tadpole embryo. Thousands of them will lie here for around three weeks until they hatch into wriggling tadpoles. Frogs are highly negligent parents and, consequently, from the thousands of eggs laid in shallow warmish water, only a small proportion will reach frog adulthood. You can sometimes find frogspawn in puddles. Frogs do spawn here, but it is a risky manoeuvre as puddles can evaporate leaving the spawn high and dry.

LESSER CELANDINE

If you see a buttercup in early spring it will be the lesser celandine, cousin to the meadow buttercups of summer. The flowers are a joyous yellow, with green or gold streaks on the outer sides. Unless you are a lover of striped lawns, this is a very welcome burst of colour at an otherwise slightly cold and grey time of year. Lesser Celandine spreads rampantly, but when it is spreading along hedgerows, ditches and shady footpaths, who cares? It was William Wordsworth's favourite flower, and he wrote three poems about it, partly to make up for the fact that until then it had not appeared in English verse.

MOLE

Although they are one of the most common mammals in Britain you rarely get to see a mole, but you can certainly spot where they have been. Molehills are common on golf courses, parks and even lawns. Moles are avid tunnellers living almost completely underground. In very dry weather, you may see one in the early morning attacking a clump of grass searching for worms and beetles. They look very much like Mole from *The Wind in the Willows* minus, of course, the spectacles and waistcoat: short and tubby with no neck to speak of, they are covered in velvety, black fur and have broad front paws with strong claws. Outside the mating season they live very solitary lives, each mole sticking to its own tunnel system.

9 MARCH
WEEDS

What is a weed? Famously it is any plant growing in the wrong place, which is a very subjective answer, but a charming wild flower may be regarded as an invasive and unwelcome weed if it crosses the garden fence or grows in a crack in the pavement. Each gardener knows exactly which plants they regard as weeds, but the definition may vary depending upon the type of garden. These are the survivors, the opportunists of the horticultural world. The next time you see a patch of 'weeds', look at the individual plants more closely, you may find more beauty than you expected.

10 MARCH
CROW

The crow is definitely a bird that could do with some help from a good PR agency. Crows have long been associated with death or seen as the bearer of bad tidings, and a flock is known as a murder of crows. They are however one of the most intelligent of birds with an intellect believed to be on a par with that of chimpanzees. Crows use simple tools, have an excellent memory and are thought to be able to recognize faces. Resourceful scavengers, they will feed on almost anything and are frequent visitors to unguarded rubbish bins.

Crows look similar to ravens and rooks. One can theoretically distinguish them by size, with ravens the largest (but this is not very helpful if you don't have them lined up, and obviously a baby raven is smaller than an adult crow). A better guide is the beak. Ravens have larger, more hooked bills, the crow's is quite short and that of the rook is a greyish white. Ravens and crows have much sleeker feathers too; the rook is shaggier with the appearance of wearing rather charming short feathered trousers.

HART'S TONGUE FERN

Most ferns look very like bracken with deeply divided leaves, and you can probably get away with calling them either 'ferns' or 'bracken' fairly indiscriminately. Hart's tongue fern looks very different and, don't worry, you won't need to find a hart and look at its tongue to be able to identify one. They grow in rosettes with each strap-shaped leaf uncurling in spring to form a tongue about 50 centimetres (20 inches) long. The pale green, leathery leaves are evergreen with conspicuous stripes of orange spores on their undersides. They like damp woodlands but will also grow in brickwork, where they usually form miniature versions of themselves. The Victorians had a craze for ferns, and this was one of the most popular species at the time.

YOUNG LEAVES OF HAWTHORN

Hawthorn means 'hedge thorn', but in urban areas you are more likely to see a hawthorn in its full glory as a tree, with twisted stems, deeply fissured bark and, of course, thorns. It is also known as the bread and cheese tree, as the young leaves, buds and flowers, which emerge in spring, are all edible. We are not convinced that hawthorn will ever become a replacement for the traditional ploughman's lunch, but it is a tasty and interesting addition to salads. Obviously, forage well away from busy roads and polluted areas.

13 MARCH

WHAT IS A PEST?

From the human point of view, there is a fine balance between a species being interesting and welcome or a pest. Pigeons are regarded as 'rats with wings' while doves get much better press. Grey squirrels are reviled for 'pushing out' the admittedly cuter (and native) red variety, but they are simply trying to survive. Rats are definitely pests, but they are also extremely interesting and the owls who eat them would, if they thought about such things, regard them as useful. The trick is not to look at nature solely from our point of view; from the point of view of the natural world, humans may be the greatest pest of all.

RIVER CLEAN-UP

There is nothing like a good spring clean. Spring is an ideal time for
a river clean-up: water levels start to drop and rubbish is easier to
spot. Taking care of waterways is an important part of maintaining
a healthy and balanced environment and protecting local flora and
fauna. The Canal & River Trust organizes regular clean-up days –
look on their website (canalrivertrust.org.uk) to find out how to
volunteer. Local community groups will also often run clean-up days
in their area.

15 MARCH

BUMBLEBEE AND BEE

You cannot help but love bumblebees and bees. Even the sting –
and they don't sting unless provoked – is no deterrent. There is
something inherently appealing in that fat, fuzzy shape. It helps, of
course, that bees produce delicious honey and are powerful symbols
of industry, wealth and prosperity, which is one reason many early
coins carried a picture of a bee.

   Determined apiarists have hives everywhere, in allotments
and palaces, even on rooftops. Some say that the urban environment
is healthier for bees, as the wide variety of plants in city gardens,
parks and window boxes provides a richer and more varied diet. For
apiarists, their bees become part of the family and there is a long
tradition of telling the bees of important events, particularly births
and deaths. When a beekeeper dies, someone must go to the hive
drape it with a black ribbon and say to the bees, 'Your mistress (or
master) is dead, but don't you go. Your new master will be a good
master to you.' This is supposed to ensure that the bees do not stop
producing honey or abandon the hive altogether.

16 MARCH
WIND

March has always been known as a windy month. The changeable nature of weather at this time of year, when hot air to the south meets cold air to the north, builds pressure and creates winds. Winds vary from a light breeze to full-scale hurricane force, and accordingly can merely cool you down or do serious damage. Winds are named for where they come from, not where they are going, so a west wind blows from the west. Only one British wind has its own proper name, the Helm Wind, a north-easterly on Cross Fell in Cumbria. Storms are a different matter, as only those likely to cause significant damage are named.

17 MARCH
CLOVER: THE SHAMROCK

The shamrock or little clover is the flower of Ireland. It was originally used as a charm against witches and faeries, but St Patrick used the three leaves to explain the Trinity, shifting the symbolism of the plant from pagan to Christian. Like all clovers, the shamrock is a lucky plant, regardless of the number of leaves. St Patrick's Day is one of the many markers of the end of winter and the beginning of spring, and at times like this – when folklore tells that magic is near to the surface of the human world – everyone, regardless of their beliefs or religion, should be pleased to accept the extra luck a clover or shamrock might bring.

HEATHS AND COMMONS

Common land in Britain is unfenced land where the right to graze, take fallen wood or collect peat or bracken is granted to groups of 'commoners'. The land itself has an owner, today often the National Trust or a local authority. On many so-called commons in towns and cities the rights of commoners have fallen away (few city dwellers need a spot to graze their sheep these days), but the Countryside and Rights of Way Act 2000 has preserved a right to roam for the general public, and these spaces have become recreational areas and havens for wildlife. Grazing kept these areas as heaths, with scrubby vegetation, rough grasses or bracken and few trees, although the change in use has seen many regenerate as woodland.

ROBERT MACFARLANE, THE WILD PLACES

This may seem a surprising book to recommend in a book on urban nature, but every so often you need to look at the wider picture in order to see what is in front of you with fresh eyes. Robert Macfarlane went in search of the truly wild places in Britain and Ireland, most of which are in remote locations, far from any towns or cities, but the point is not where they are, but what he found. These are mostly unspoilt areas, and they give a fascinating picture of the natural world, which once covered all the land. It is still there and ready to re-emerge given half the chance: an unused plot of land, a canal bank or even a crack in a wall. Remember what was there before the roads and buildings and keep an eye out for it.

20 MARCH
SPRING EQUINOX

Our diaries are crammed with important events, but nature has just
four: the spring and autumn equinoxes and the summer and winter
solstices, which mark the ratio of dark and light on each day. The
spring equinox is one of the two points in the year when the day
is the same length as night. From now on, there is more and more
light each day. It doesn't matter where you live or what the weather
is doing, today marks the proper start of the move towards summer.
Most plants need light rather than warmth to grow, and this is the
moment when many begin to emerge. Insects follow the lead of the
plants, and birds and animals follow them. The whole natural world
is waking up.

21 MARCH
COMMUNITY GARDENS

Community gardens have almost as many origins as there are
gardens: car parks, failed building projects, bomb sites, quarries,
disused railways. What they all have in common is a passionate
group of supporters and volunteers who tend the gardens and
fight, if necessary, for their survival. As the name suggests, they
are as much a place to meet as a garden, and many are not even
gardens in the traditional sense but more of a nature reserve; a place
where children can learn about wildlife as well as gardening. Some
are open all year round; others on particular days as part of the
National Garden Scheme.

22 MARCH
PRIMROSE

These dainty, pale yellow flowers appear early in the year, growing
on single stems from a base rosette of crinkly, green leaves. Cowslips
and oxlips, similar at first glance, flower later and hold their flowers
in clusters rather than singly. The name 'primrose' comes from the
Latin *prima rosa* or 'first rose', and although these plants reach barely
a few centimetres above the ground, the flowers do resemble the
simple wild rose. For much of the 20th century they were one of the
most popular flowers to pick, resulting in increased rarity around
cities. They like growing on banks, either beside roads or along
railway lines, which, helpfully for us, raises the plants up and makes
them easier to spot.

23 MARCH
STREET TREES

Most street trees have a fairly tough life; they may look healthy,
but their roots face a constant battle trying to spread out in the
compacted soil beneath our roads and pavements. Roadworks disturb
their roots further, and above ground they suffer the unwelcome
markings of dogs, and are often pruned to suit the needs of the street
rather than the tree. These trees produce oxygen, cool our streets
and lift our hearts with their beauty. In return, give your tree a drink
– a bucket of water every two weeks can make all the difference in
a hot summer. Young trees especially need our help; check the ties
don't get too tight as they grow, and gently pull away any weeds that
will compete with the tree. The local council may plant the trees, but
we should all help look after them.

HERON

The long-legged heron, Britain's tallest bird, is an avian supermodel.
Whether it is standing on one leg perfecting its own version of tai
chi, or flying gracefully through the air with those long legs trailing
behind, the heron always looks poised and elegant. The grey heron
is the one you are most likely to see around urban ponds, fountains
or waterways. A patient fisher, a heron will stand stock-still by the
water waiting to spot a potential snack, or occasionally stalk its prey
through the shallows. Its white face is distinguished by a long, black
'eyebrow', which flows into a black crest. Herons build their nests
high up in trees and will return to the same nest year after year,
making casual home improvements.

URBAN FARMS

The urban farm movement started in the 1970s, and city farms can be either a local community food production enterprise or a way of connecting city dwellers with the wider farming community, sometimes both. They have a greater importance than a day out for the children or an opportunity for a bit of pig tickling (really, you can do this). They serve as a green oasis in an otherwise highly urban area where animals, vegetable plots, food markets and composting are gathered together and help connect the local community, explore the interaction between nature and society and provide a valuable opportunity for knowledge exchange. For children brought up within cities, the urban farm demonstrates the effort and skill involved in food production, the importance of sustainability and the minimization of waste.

PUSSY WILLOW

These small, bushy trees are also called great sallow or goat willow, but for us the furry male catkins will forever make it pussy willow. They appear before the leaves and turn fluffy and yellow with pollen. The female catkins are longer and green. Unusually for a willow, the leaves are broad rather than long and narrow, but the plant hybridizes with other willows resulting in many indeterminate forms. Make the most of the fluffy catkins now and don't worry about the plant later on in the year when it is perfectly pleasant but less charming.

DAYLIGHT SAVING: SPRING FORWARD

If you are never able to remember which way the clocks go at this time of year, the saying 'Spring forward, fall back' can help. We like this because it has us looking forward with enthusiasm rather than moaning that we've lost an hour in bed. Change your clocks the night before and you'll barely notice the difference. Benjamin Franklin suggested the idea in 1784 as a way to conserve candles, but as far as we are concerned it gives us an extra hour of light in the evening. You can go for a walk, do some gardening or simply look out of the window. Most of us are rushing around in the morning, with little time to notice whether it is light or dark, but an extra hour of light at the end of the day makes a huge difference. Make the most of it.

CANALS AND WATERWAYS

Canals are a legacy of Britain's industrial past when they served as a vital transport network. The main industrial cities of the north are particularly well-endowed. Today their function is more recreational, and the work of the Canal & River Trust on cleaning up disused canals means that whether you chose to explore by narrowboat or towpath, these waterways are a rich source of opportunities for wildlife spotting. Like many places, it is interesting to examine canals at different times of the year, and see how this affects what you can see (your wildlife diary will be invaluable here). Look all around: the trees and plants alongside the river, bird and insect life, peer into the water for fish and, if you are lucky, a water vole.

canalrivertrust.org.uk

HEDGEHOG

Now that spring has taken hold, the hedgehog at the bottom of
the garden has woken up and poked out his (or her, we are a little
fuzzy on hedgehog sexing) snout to begin their quest for a post-
hibernation snack. After a three-month kip, they are peckish and
need to build up their strength for the mating season to come.
Hedgehogs roam quite large distances in search of food, and it
is important to make a small hole (14 x 14 centimetres [5½ x 5½
inches] is enough) in any fences so they can travel. Raised on Mrs
Tiggy-Winkle, we find hedgehogs tremendously appealing: their
furry faces, beady little eyes and snuffling noises. And, of course,
their taste in slugs is a boon for the keen gardener.

ADDER

It's unusual but not unheard of for an adder to slither across the path
as we cross the common. Adders love rough grassy spots. Although
they are Britain's only venomous snake, they are really very timid and
will only bite as an absolute last resort, and even then it is unlikely
to be fatal although you should seek immediate medical advice. Male
adders are white or pale grey with a distinctive black zigzag down the
length of their bodies, while females have similar markings but in
brown and darker brown. Adders shed their skin in spring when they
come out of hibernation and have 'grown out' of their old one. In
the past, a shed skin was regarded as a charm against infection and
worn as a hatband. Adders are the most abstemious of eaters, usually
only eating 6–10 times a year.

They are great sunbathers, particularly when they first come out
of hibernation. They also have their own fan club, Adders Are Amazing! –
check them and other reptile fan groups out on arguk.org.

NETTLE

Nettles sting, we all know that, but butterflies love them and they
can be made into nutritious liquid fertilizer for your garden. They
are delicious to eat (cook them first), a bit like good spinach and are
packed with vitamin C, iron and protein. The smaller leaves at the top
are the tastiest, but you must not eat nettles once the flowers appear
as the plant becomes harmful. They have tall, purple-green stems and
shield-shaped leaves covered with fine hairs, which are what produces
the sting. Supposedly, if you grasp a nettle firmly it won't sting you,
but we don't advise you to try this as some other nearby leaf is likely to
gently brush against (and sting) you when you are not looking.

APRIL

By April, there is so much to see we are spoilt for choice. There are fox cubs; the swallows are back from their winter holidays; the frogspawn has become tadpoles; and the fat, sticky horse chestnut buds burst forth with an energy that must be seen to be believed.

If you are in London, Kew Gardens has an amazing bluebell wood, which is well worth a trip. It may still be a bit chilly, but that is what cardigans are for. Pick up your copy of *The Yellow Book* and plan out which gardens you will visit this year.

The trees along this city street,
Save for the traffic and the trains,
Would make a sound as thin and sweet
As trees in country lanes.

From 'City Trees' by Edna St Vincent Millay

1 APRIL
URBAN MYTHS

During the cold, dark days, we have been known to amuse ourselves
by examining some of those urban nature myths to see what is
behind them. There are plenty: crocodiles in the Thames; the Hull
Hell Cat; shoals of goldfish the size of dinner plates swimming
through sewers. Most are quickly debunked, while others persist
for years. The crocodile, despite a convincing photograph, turned
out to be a garden ornament. Big cat sightings are common and it
is possible that escapees from private collections have naturalized.
Just occasionally there are some once-in-a-lifetime spots: lost whales
that have ventured up-river; the boa constrictor that turned up in
a McDonalds in Bognor Regis; or a red squirrel in an urban park.
Mostly though, the usual rule of thumb applies – if it sounds too
amazing to be true, it probably is.

2 APRIL
GREY SQUIRREL

Sitting in the garden there is a mighty thump as a plump squirrel
lands on the roof of the shed: a surprise as much for the squirrel
as for us. Usually, these dainty little creatures are amazing acrobats
with padded feet that enable them to jump and land safely from
6 metres (20 feet) up, and to scamper along the thinnest branch.
Grey squirrels were introduced into Britain in the 19th century and
the Duke of Bedford, a keen squirrel lover, gifted them around the
country hastening their spread. Voracious eaters, they are constantly
nibbling on a titbit held between their front paws (nuts and bulbs
are particular favourites). As their teeth never stop growing, this is as
much to keep that buck-toothiness at bay as it is for nourishment.

GUERRILLA GARDENING

Guerrilla gardening covers everything from putting a single plant just outside your boundary fence to taking over a huge disused plot. It is equally loved and hated and is done at night for secrecy or by day for publicity. Some people plant food, others flowers. In recent years, around the world, bus stops have bloomed, roundabouts have produced unexpected harvests and derelict land has blossomed. The easiest way to guerrilla garden is to throw seed bombs. If you are planting garden plants, remember they need to be tough enough to survive largely untended. City councils vary widely in their attitudes; we have gone back to some 'gardens' and found they have been flattened with weed killer, while other councils offer help and support. Guerrilla gardening can be immensely satisfying, but we recommend you do a little research first.

guerrillagardening.org
kabloom.co.uk
wildlifetrusts.org

4 APRIL
CHERRY BLOSSOM

The arrival of spring blossom is one of the most important markers in the natural world's calendar, marking the true end of winter. From a human perspective, whose heart is not lifted by the sight of pink and white flowers turning trees into giant ice cream cones or low-flying, magical, pink-tinged clouds? There is a great variety of ornamental and fruiting cherry trees, meaning you can see blossom from mid-winter to late spring, but it is now at its peak with plum, pear and apple blossom extending the spectacle in colours ranging from pure white to a splendidly brash cerise. You can see cherry trees in orchards, but they also make good street trees, so look out for sudden bursts of colour in estates, public parks and private gardens.

5 APRIL
WATER VOLE

Most of you have probably met a water vole – Ratty in *The Wind in the Willows*. He is called a water rat, but water voles have many names: water rattan, water mole, craber, water dog, earth hound and water campagnol. They are difficult to spot, but if you see a 'rat' near a stream or river, look again. Water voles have furry tails, tiny furry ears, beady black eyes and very sharp, yellowy orange teeth. They were living in Britain over 9,000 years ago, but by 1998 water voles were in danger of disappearing altogether, mostly due to American mink, pollution and riverside development. Laws have now been passed to protect water voles, and good places to see them are wetland sites such as those run by the WWT (Wildfowl & Wetlands Trust).

GARDENING WHEREVER YOU CAN

Having a garden in an urban area is a luxury. Perhaps you have
had that moment in conversation when someone who lives in the
countryside tells you that they only have a small garden, a mere half-
acre. You laugh hollowly and reply you have a mere half-metre. But
no matter, any outside space can be used to grow plants. All they
need is a little light; you can provide the container, potting compost,
food and water. Look outside your house or flat; is there really no
space for a single plant? In one pot you could grow food, a climber
round the door or flowers to attract bees. Whatever you grow it will
be a patch of nature and, if you move, you can simply take the pot
with you.

CENTIPEDE AND MILLIPEDE

As we turn over a stone there is a quick wriggle of centipedes (or
was that millipedes) fleeing in multiple directions. These little
arthropods are nearly blind and are most comfortable in the
dark. And they are quick. There is no chance of having them hang
around while we count their legs, although to be honest this may
not help: centipedes can, in fact, have up to 350. If you need to
tell them apart, centipedes are the flatter ones, with one set of legs
per segment, while millipedes are more cylindrical, their legs stick
out and they have two pairs per segment. Millipedes are benign
vegetarians feeding on decaying plants, while the aggressive centipede
kills its lunch (silverfish, spiders, cockroaches) with a venomous bite.

OAK LEAVES

Oak leaves and acorns are two of the most easily identifiable finds in nature. Oak trees also stand out – imposing structures in any park or garden. There are two oak trees native to Britain: the English, pedunculate or common, and the sessile or durmast. They look similar, but the leaves of the sessile oak are stalked and the acorns stalkless. Those of the English oak are vice versa. If this sounds like something out of *1066 and All That*, don't worry too much – the two frequently hybridize, allowing you to simply say, 'That's an oak,' in an authoritative tone. Historically important for building houses and ships, oaks are home to more than 300 species of insects and, in 1651, provided a perfect hiding place for the future King Charles II.

WREN

We are not sure we have ever met a wren called Jenny, but we do have one that is a regular visitor to the bird feeder. You are quite likely to be able to spot a wren, they are widespread, and, in fact, the wren is the most common breeding bird in Britain. It is the little bird with the big voice; its loud, bubbly song a delight. Smaller even than a dunnock, the wren is almost round with a short tail that tilts up. As the tiniest bird in England, Edward VIII had the wren put on the reverse of the farthing, the smallest denomination coin of the realm at the time. In cold weather little wrens will nest communally, with several snuggling in together in the same nest box.

DANDELION

Children use dandelions to tell the time by blowing away the fluffy seed heads, and each tiny seed that lands somewhere suitable will grow into a thistly little plant. Dandelions are not fussy: cracks in pavements, grassy paths or unattended flower beds all make perfect homes. Each flower looks like a miniature sun with many yellow florets (tiny flowers) packed closely together. The name comes from the French *dent-de-lion* or 'lion's-tooth' possibly from the jagged shape of the leaves. These can be eaten in salads, the flowers made into wine and the roots used to make a coffee substitute. Don't be tempted to pick the flowers; they need the outdoor sunshine and will close up indoors.

ANT

Like bees and horses, historically ants have had a good and well-deserved reputation as industrious workers. There is something mesmerizing about watching a line of ants trailing back to the nest carrying food to lay before their queen. The strongest animal in proportion to its size, a single ant can carry up to 50 times its own body weight, and they sometimes work in teams to carry Herculean loads. And viva girl power! All those ants are female: the males don't work and are produced only at mating season for their brief moment of usefulness. The queen herself can live for years; she mates only on that one day, but can store sperm in her abdomen to release at intervals during her lifetime.

ORCHARDS FOR ALL LIFE

Traditionally, orchards were planted to provide fruit: apples, pears, cherries, quinces and the like. Nearly all commercial orchards have vanished from cities, but in their place community orchards are growing up thanks to organizations such as The Orchard Project. Old orchards are being revived and new ones planted. These orchards, however small, provide a valuable focus for local people and an even more valuable home for insects, animals, birds and wild flowers. Find an orchard and try to visit it throughout the year, enjoying its blossom, harvest and a host of other activities.

NEWT

Newts are lizards' amphibian cousins. We quite often see them on late spring evenings in the park near the pond. Although newts breed and lay their eggs in water, they actually spend most of their lives on land in dark, dampish spots, such as under stones or logs or in thick grass hiding away from the sun, and coming out at night to feed. During the winter they hibernate completely. They have soft, damp skin and some varieties are a little warty. During the breeding season the male common newt goes all out and makes an effort, dressing to impress. He develops an orange belly and throat with darker spots and grows a crest down his back.

COWSLIPS AND OXLIPS

These grow in clumps of small, dainty, yellow flowers that nod attractively in the breeze. The name comes from the belief that the plant would grow in fields grazed by cows or oxen, but we've found them at the edges of paths and parks where no cows have been. They both look similar to primroses but there are clear differences: primrose flowers grow singly on a woolly stem, while oxlips and cowslips grow in clusters. Oxlip flowers are pale and flattish, while cowslips are a rich yellow with orange streaks and are bell-shaped. Oxlip leaves are less woolly and wrinkly than those of primroses. False oxlips are a hybrid of the cowslip and primrose, but this is a coarser-looking flower, with few of its parents' charms. All this really only helps if you have all four flowers lined up together, but no matter, all belong to the primrose family and all are delightful.

BIRDSONG

There is a wide diversity of birdsong, from chittering to hoots, trills to caws, warbling to screeches. To further complicate the task of identification, the same bird may make different sounds depending on need or mood, and some, like the starling, will imitate others. Identifying a bird by its song is an art, but don't be intimidated. Begin with something easy such as a gull or a cuckoo. A good starting point is listening to Radio 4's 'Tweet of the Day', which profiles a wide range of different birdsong, and the RSPB has a useful identifier on its website (rspb.org.uk). And you can just listen; you don't need to know who is singing to appreciate the song.

STICKLEBACK

We used to catch sticklebacks at the local weir as children. Their buggy eyes and gulping mouths give them such a friendly look. The male (he's the one with the bright red underside) is a solo dad. He sucks up mud and sand from the river bed making a shallow depression, and then glues together vegetation with excreted mucus to build a dome-like nest (a kind of piscine igloo if you will). He then coaxes (or chases) a female inside where she lays her eggs, and he follows in to fertilize them. Not being a monogamous fish, he repeats this process until his nest is full. He will then guard the eggs and raise the young himself.

BLACKBIRD

The blackbird is a noisy little chap with a fluting, carefree song, who loves feeding on lawns and flower beds. If you see a blackbird with his head cocked on one side he is listening for earthworms.

The male blackbird is the one you recognize: all black save for a yellow ring around his eye and a bright yellow bill. The female is a dark brown with streaks on her chest and throat.

Numbers swell during winter when native blackbirds are joined by migrants from Scandinavia and the Baltic states. Blackbirds also glory in having their own patron saint, Saint Kevin of Glendalough, an Irish monk who was so caught up in his prayers that a blackbird came down and made a nest in his hand.

18 APRIL

MACKEREL SKY

The mackerel sky is one our favourite cloud formations. There is something so amazingly tactile about the undulating ripples of cirrocumulus clouds, you just want to reach out and run your fingers over the bumps. The name comes from the clouds' resemblance to fish scales, although we have always thought they look more like the ripples on sand after the tide recedes or possibly a shoal of fast-moving fish. On the continent, they call it sheep's clouds as they see flocks of woolly sheep. A mackerel sky is supposed to signify a change in the weather, often running half a day ahead of a short burst of rain. 'Mackerel sky, mackerel sky, never long wet, never long dry.'

DAISY

Lawns spotted with daisies or regimented in pristine stripes? Like miniskirts and so much else, daisies come and go in fashion. We remember (as children, but we'd happily do it now too) sitting on the grass making daisy chains, either in competition to see who could make the longest, or as a production line creating free necklaces and bracelets. Like many wild flowers, daisies are survivors; the mower may chop their heads off but, unlike the unfortunate French aristocracy, the resilient little plant will soon produce more flowers. Their name comes from the Old English *daes eage* meaning 'day's eye' – they open with the morning sunlight and close in the evening, revealing their pink-tinged undersides. Sensibly they also close when it is cloudy.

SLUG

When weeping over destroyed seedlings, we have been known to comment that the slug is the creature only its mother can love, but they are really rather amazing. A single slug can have up to 27,000 teeth, so no wonder it makes short work of that lettuce. The best time for a bit of slug spotting is early morning after rain. Look closely as there are around 40 different species: round ones, long ones, fat ones, thin ones, red ones, brown ones. A single garden can harbour up to 20,000 slugs on a bad day. Most slugs have two sets of feelers: an upper, light-sensing pair (their eyes if you will), and a bottom pair providing their sense of smell. They 'breathe' through opening and shutting a hole in their sides.

DUNNOCK

Dunnocks are one of the most common garden birds. You have almost certainly seen one, although you may have thought it a sparrow, as they look very similar. Their other common name is hedge sparrow. These shy little brown and grey birds like to spend a lot of time on the ground, shuffling about in flower beds and near bushes, although things can get heated if two rival males meet, engendering much name-calling and flicking of wing tips. Female dunnocks may mate with more than one male and the resultant paternal confusion means both end up supplying her chicks with food.

EARTH DAY

The first Earth Day was held in America in 1970, the brainchild of Senator Gaylord Nelson from Wisconsin, who wanted to use the energy of the student anti-war movement to raise awareness of air and water pollution. The event marks the start of the modern environmental movement. More than a billion people worldwide now observe Earth Day, fighting to change human behaviour and achieve a clean environment. Events are either held at the spring equinox or on or around 22 April. Details of how to take part in an event, organize one or raise awareness in your area can be found on the Earth Day website.

earthday.org

ST GEORGE'S MUSHROOM

St George's mushroom is so named because it appears in the spring around St George's Day (23 April) and is usually the first mushroom to appear, heralding the start of the season. It is a very pale mushroom with a creamy white cap and gills, and in towns can be spotted on verges and by the sides of roads, either scattered or sometimes in fairy rings. The best time to look is just after a shower of rain. They are described as smelling mealy or like wet flour. The St George's mushroom is one of our edible fungi, but please do be very, very careful when picking mushrooms to eat and only proceed with the advice of an expert as mistakes can be fatal.

24 APRIL

SPLASH OR SOAK

Oak before Ash, in for a splash.
Ash before Oak, in for a soak.

There are two things to be aware of about this old rhyme: the first is that neither combination assures us a dry summer; and second that a German rhyme says the exact opposite. Oak and ash trees come into leaf at roughly the same time (March to May), but while the oak is dependent on warm temperatures, ash trees come into leaf as the hours of daylight lengthen. A warm spring will put the oak in the lead while a cold one will give the ash the advantage. Warm springs are sometimes followed by dry summers but not always. So much for ancient lore.

25 APRIL

SWALLOW

If there is a line of birds on a telegraph wire, swallows are a safe
bet. Swifts, while looking similar, never roost. Swallows have much
longer tails too, and more white on their bodies. Swallows overwinter
in South Africa, returning to Britain in April and May and staying
until October. They like to nest on beams and ledges, and you can
sometimes see their nests, which are twiggy mud cups, under railway
arches or in church porches. Swallows will reuse the same nests year
after year. This makes total sense to us; if you have just flown 10,000
kilometres (6,200 miles) to your breeding ground, the last thing you
will feel like doing is a spot of home-building.

GRASS SNAKE

The grey-green grass snake (try saying that quickly after a pint or two) is our longest snake, growing to over 1 metre (3¼ feet). As befits the name, they hang out in long grass, ideally close to water, and can be quite fond of garden ponds. They lay their eggs in warm piles of rotting vegetation (yes, that does mean your compost heap). But fear not, they are not venomous and are more frightened of you than you are of them. Shy by nature, if disturbed a grass snake will slither off or, if there is no clear exit, play doggo rolling onto its back with its tongue hanging out – significantly less cute than when done by a puppy, but amusing nevertheless.

BLUEBELL

Jane grew up near a bluebell wood and shamelessly took it for granted, assuming that all woods had carpets of blue in spring. Sally soon corrected her. It is worth hunting out a bluebell wood at this time of year; the effect is magical both from a distance and when you look at the flowers closely (the National Trust, the Woodland Trust and local websites all list good sites). Don't be tempted to pick the flowers to take home; it is against the law (and they don't last well). Native bluebells have narrow, dark blue flowers that droop daintily to one side. Like Spanish sailors in Elizabethan times, Spanish bluebells are trying to invade the country. They are paler, more upright and, sadly, more robust. Gum from the sap was used to bind books and attach feathers onto arrows, but a more charming legend is that the bells were used by fairies to trap passers-by.

28 APRIL

FOX CUB

Foxes have litters in March, but you must wait to spot cubs as they spend the first few weeks of their lives underground. If by any chance a vixen has made a den under your shed, please do keep your distance. The vixen will stay with the cubs almost full-time when they are very young as they are blind and vulnerable. Other foxes will bring food, so there may be lots of foxy business going on which you can watch from afar.

By late April they will be venturing out and you may be able to spot them playing on railway embankments and patches of waste ground. We once saw a bold family sunbathing on the flat roof of a neighbour's extension.

THE YELLOW BOOK

Properly titled *The Garden Visitor's Handbook*, this invaluable guide is published every year in a brilliant yellow book (you can also find the information on the National Gardens Scheme website but 'The Website With Yellow Edges' wasn't such a catchy title). The book and website list gardens, public and private, large and small, which are open in aid of charity on particular days. It is a brilliant way to see private gardens and pick up ideas; the gardeners are always on hand to answer questions and offer advice and many also serve delicious teas and sell plants. The gardeners choose when to open so you can be sure of seeing the gardens at their best, so there is none of that, 'Oh, you should have been here last week.'

CHAFFINCH

Chaffinches have it all, being both pretty and musical. A noted songster, the Victorians used to keep chaffinches for their powerful voice and held singing matches between cock birds. The males have pinky coloured breasts and cheeks with a blue-grey head and back, which tapers to a grey-green, and white flashes across the wings; the females are a slightly duller browny grey. The old country name for the chaffinch was the bachelor bird as all-male flocks can often be seen, particularly during winter. The name 'chaffinch' refers to their feeding on the chaff left over from threshing, or in the case of the city birds, the chaff that fell out of working horses' nosebags. They like to feed on the ground and are more comfortable with the crumbs from the bird table than in visiting the table itself.

MAY

The natural world is out celebrating in May and we should too.
Plants are clothed in vibrant green, giving a unique light to the
month and the birds and animals born earlier in spring are taking
their first tentative steps, flaps or paddles. The rhyme 'gathering
nuts in May' may seem odd as nuts are harvested in the autumn,
but it actually means 'knots' rather than 'nuts' – knots or bunches
of flowers for May Day festivals. Find a festival, make a posy and
enjoy spring at its height.

———————

The trees are coming into leaf
Like something almost being said;

From 'The Trees' by Philip Larkin

DEW

Dew has the ability to scatter the earth with diamonds – tiny drops of water that cling to any and every surface on certain magical mornings. Most of the water that rises from the earth's surface is held in the air as a gas called water vapour. The dew point is the temperature at which the water vapour condenses and turns back into water. The Druids rightly regarded moisture produced by nature as sacred and gave thanks on May Day for its life-giving properties. Apparently, for a flawless complexion, you should wash your face in the dew on May morning, but we have no proof of this.

2 MAY
HORSE CHESTNUT BLOSSOM

All spring blossom is fabulous, but there is something very special about the flowers of the horse chestnut tree. These are huge, majestic trees and every year, at about this time, nature decorates them with hundreds of glorious candles. Large, sticky buds grow into upright spires of white flowers splashed inside with pink. Red horse chestnuts have (in our opinion) even more wonderful, deep, pinky red flowers, which stand out against the dark green foliage. Close up the flowers don't smell great, but enjoy them from a distance. Horse chestnuts secrete chemicals to prevent competition growing nearby so the trees usually stand in stately glory on their own.

MOHICANS IN THE PARK: NO-MOW MAY

From May onwards, you may notice your local park becoming shaggier. This is not laziness on the part of the gardeners, but is down to No-Mow May, a brilliant scheme whereby grass is allowed to grow longer, and wild flowers to flourish without having their heads chopped off. Forget traditional lawns with stripes; research shows that the best look for wildlife is the Mohican, with some grass uncut and other areas cut about once a month. This allows slow-growing flowers, such as field scabious and knapweed, to flourish in the uncut areas while the nectar-rich daisies and white clover will produce more blooms after being cut. The pollinating insects will be happier, the flowers will be happier and there will be so much more to look at than a simple stripe – so we should be happier too.

OAK PROCESSIONARY MOTH

This is definitely one of those conflicted moments. Watching a line of hairy, white caterpillars snake down an oak trunk in a nose-to-tail caravan is fascinating, they look so purposeful. There is a clear leader out in front and rows of soldiers following behind in arrow formation. Higher up in the tree we can see a nest made from white webbing, a bit like a teardrop-shaped spider's nest. However, this is not great news for the noble oak, as these very hungry caterpillars chew on the leaves and cause serious damage to the tree. Sightings of oak processionary moth should be reported to the Forestry Commission via the online Tree Alert form. And don't get too close; the hairs on the caterpillars can cause rashes or exacerbate asthma.

MARCH WINDS AND APRIL SHOWERS BRING FORTH MAY FLOWERS

Like many weather-related proverbs, there is logic behind this old English saying, which applies in rural and urban areas alike. When the sun begins to warm the earth after winter, warm and cold air masses develop, bumping against one another and causing swirling, gusting winds. As the seasons progress, the sun rises higher in the sky, the ground begins to heat up more and the lower, warmer air rises and forms clouds when it meets the much colder air higher in the atmosphere. This is why April days can start warm and sunny, develop showers in the afternoon and then become clear, chilly nights. By May the soil is warm and damp and the days are longer – a perfect combination for hardy British flowers.

6 MAY
WILD GARLIC

The chances are you will smell ramsoms or wild garlic at the same moment that you see it. From a distance, the plants look a little like poisonous lily of the valley, but close up they are easy to tell apart. Ramsoms have starry, pure white flowers, while those of lily of the valley look like bells hanging along the stalk, plus ramsoms smell distinctly of garlic. The young leaves and flowers are excellent to eat (avoid the roots, which are harmful) but be sure to pick from places where there are no dogs. Preferring ancient woodlands and woodland streams, ramsoms are not common urban plants, but can be found in cemeteries and wooded parks, and it is definitely worth hunting them out if you are a keen cook.

7 MAY
DAWN CHORUS

The first Sunday in May is Dawn Chorus Day. Clearly no one has told the birds the timetable, as they are up and tweeting early most mornings from March to July. And who thought it was a good idea to celebrate a dawn event on a Sunday, which is surely the day of the lie-in? However, International Dawn Chorus Day, which started in Birmingham in the 1980s, is now celebrated worldwide. So, set the alarm (or leave your window open) and see which bird's song you can make out: song thrush, blackbird, robin or the warbling of the chiff chaff. The fun kicks off about an hour before sunrise.

NOBLE FALSE WIDOW SPIDER

A colleague bought in a picture on his phone of a noble false widow spider he had found in his flat. Although they look rather (okay, very) scary they are not dangerous. They can bite, but rarely do unless provoked and the bite is irritating rather than fatal. That said, the spider that bit Peter Parker, aka Spiderman, was a noble false widow, albeit a radioactive one. Their dark, shiny bodies look like enamelled jewels decorated with a skull-shaped pattern. They have occurred in Britain since the 1870s, when they arrived as stowaways on ships from Madeira and the Canary Islands. Since the 1980s, the rising temperatures that suit them have caused something of a population boom, making them one of the most common urban spiders.

MALLARD

These are the largest common ducks and can be seen on most park ponds. The male is distinctive, with a glossy, green-blue head and neck, a smart white neck band and yellow bill. The females have orange bills, and their feathers are every shade of brown you could imagine. This camouflages them beautifully so they can look after their eggs and chicks while the drakes strut about. Both have dark blue and white patches on their upper wings. Sadly, male mallards exhibit some appalling behaviour; although they sweetly pair up with a mate in winter, later in the season they will also try to force their attentions on unattached females.

WILD ABOUT GARDENS CAMPAIGN

This is a brilliant collaboration between the Wildlife Trusts and the Royal Horticultural Society to show us all how easy and satisfying it is to attract more wildlife into our gardens, regardless of their size. Creatures such as hedgehogs, house sparrows, many beetles and starlings are under threat in urban areas, and we can all help. The website has useful guides on different aspects of wildlife conservation in gardens, and each year the campaign chooses a theme to focus on. There is also a monthly newsletter. It is a sign of how much can be achieved when organizations work together and unite like-minded people.

wildaboutgardens.org.uk

LEAF COLOUR IN SPRING

Autumn is the traditional time to admire the colour of leaves, but take the opportunity to also look in spring; you will be rewarded with a thousand different shades of green. These are not the dark greens of high summer or the solid walls of evergreens, but the pale, luminous greens that light up spring. Many trees, such as blackthorn, bear their blossom on bare stems and you will notice the pink or white flowers developing a greenish tinge as the leaves grow through the flowers. Leaves grow from buds, which have remained dormant on the tree throughout the winter, and burst open in the spring sunshine. As the weather warms up, the sap rises in the tree, taking sugar and energy up to the leaf buds. The chemical chlorophyll makes the leaves particularly bright in spring.

THRUSH

The distinctive thrush, with its speckled breast, is an easy bird to identify but definitely warrants closer observation, as its beautiful, dappled brown breast markings are individual. On song thrushes, these can resemble open hearts or arrows. The largest thrush is the mistle thrush, named after its fondness for mistletoe berries, the common song thrush comes next, while the smallest is the redwing, which has a bright red smear under each wing.

Thrushes are extremely territorial and will vigorously defend food sources, like our holly tree, as well as their own nests. They feed on berries, worms, insects and even snails, which they will bash against stones to break open the shell.

THE SHAPE OF TREES

Trees come in all sorts of shapes and sizes, and you don't need to know what type of tree it is to appreciate the silhouette. That said, the shape can be a big clue in identifying the tree. Trees can be columns or cylinders acting as pointers to the sky, they can weep down over water, form broad, triangular pyramids or round lollipops on sticks. Revisit the same tree in summer and winter and compare the shape formed when the tree is clothed in leaves with the stark outline of its bare branches. When thinking about the shape of individual trees, look at how the branches grow: do they shoot straight up, standing to attention; do they lazily wander this way and that; or are they disciplined by the municipal park-keeper to form an arch?

14 MAY
SORREL

These are the tall, rusty spires of waste lands, rising to between 30 and 80 centimetres (12 and 30 inches). The branched, dark red flowers appear in early summer, the colour spreading throughout the whole plant as autumn approaches. The leaves are edible, with a sharp taste, and go well in salads, soups and fish dishes. The name 'sorrel' comes from the Old French *surele,* coming in turn from *sur* or 'sour'. According to writer and naturalist Geoffrey Grigson, the juice can be used to take rust marks out of linen, a remedy we have yet to try.

15 MAY
HERB ROBERT

This is a wild cranesbill geranium, a smaller version of its garden cousins. The stems are an eye-catching red and the leaves, also tinged with red, stink when you crush them – earning it the other common names stinking Bob or stinking cranesbill. The flowers are a pretty pink, so our advice is to look at the flowers but not to touch the leaves. No one knows who poor Robert or Bob was, but the name may be a corruption of Robin, leading to Robin Hood or Shakespeare's Robin Goodfellow – Puck from *A Midsummer Night's Dream.* Potential Roberts include a Duke of Normandy and Robert of Molesme, an abbot and herbalist, but the name may simply be a corruption of *ruber,* the Latin for red. Take your pick.

WASP

An orderly queue of wasps entering and exiting the porch of a vacant house suggests there may be a wasps' nest there. Closer (but cautious) examination reveals a greyish, papery looking shape. Social wasps will build nests anywhere relatively dry and structurally sound. It is fascinating to watch the nest, but do keep a safe distance.

Over winter the queen hibernates under tree bark or in building crevices then, come spring, she will be out seeking a spot to build a colony. She lays eggs that hatch into female worker wasps who take over construction and keep her fed while she gets on with egg laying. By the end of summer, those eggs will have hatched, and the drones and new queens dispersed to hibernate and start the process once again.

FORGET-ME-NOT

Considering how small they are, forget-me-not flowers pack a lot of punch. Tiny (5 milimetres [$\frac{1}{5}$ inch] across) and low-growing, the pale blue flowers with yellow centres bloom from mid-spring to autumn, and even into winter. Traditionally, this was regarded as a flower of love – if you wore it you would not be forgotten by your loved one. There is a German tale of a knight and his lady walking by a river. He picks forget-me-nots for her, but unfortunately falls into the water; throwing the flowers to his love he cries, 'Forget-me-not', before drowning.

18 MAY
SWIFT

We know summer is coming as we welcome back the swifts from their winter stay in Africa. Common in cities as they love to nest under eaves and roof spaces, swifts never seem to land: they eat, mate and even sleep on the wing. Because we usually see them from a distance, swifts look black, but in fact (get out those binoculars and check) they are a sooty brown with a pale throat. It is their shape that is so distinctive – with their crescent-shaped wing profile and forked tail silhouetted against the sky, they are a lovely sight. Swift by name and swift by nature, they are one of the fastest birds, with flying speeds of up to 100 kilometres (70 miles) per hour.

19 MAY
AGRIMONY

You will see agrimony in any long grass, first as narrow, green spires in May when the flower is in bud, then as rich, primrose-yellow flowers in summer, and finally as spiky rust-coloured seed heads in autumn. The seeds will stick to you or any passing animal in the plant's attempt to spread as widely as possible. When you find them on your clothes, scatter them thoughtfully for the plant. It sets seed late so it suffers if the grass is cut too often or too early. It is also known as church-steeples, fairy's wand, rat's tails and, in Dorset, money-in-both-pockets. Agrimony is about 60 centimetres (2 feet) tall, while its relative fragrant agrimony is taller, scarcer and obviously more fragrant.

SCENT

Much of the natural world relies on scent, to navigate, hunt or attract pollinators. The human sense of smell is more highly developed than many of us probably realize, to the extent that if we needed to, our brains could navigate by smell. We all know the smell of a rose, or mown grass and, of course, in cities, the smell of petrol fumes, but next time you are in a park or on a walk in a green area try to notice the different smells. In a surprisingly short space of time, you will find your sense of smell has become more acute.

21 MAY

ROWAN TREE

The rowan is also called the mountain ash or quickbeam. It might just as well be called the urban street tree as it is one of the trees you are most likely to see growing along residential roads. The leaves are feathery and pale green, sometimes tinged with crimson later in the year and in spring there are clusters of small, white flowers; berries follow later. Traditionally it was often planted near homes as it was thought to ward off witches and its old Celtic name, *fid na ndrnad*, means 'the wizard's tree'. It is also believed to act as a doorway between worlds, which would make an ordinary suburban street much more interesting.

MOTH

Clothes moths and carpet moths are both little, brownish, almost golden-coloured moths, and are the ones you are most likely to see inside your house. Global warming, with its increased temperatures, our well-insulated, warm houses and range of natural fibres (preferred by the moth) have created perfect conditions for a population boom. Moths are certainly not endangered. A house with wool insulation is like an all-you-can-eat buffet for a moth. In fact, it's not actually the moths that do the munching, it is the larvae, but depending on conditions they can take anywhere between two months and two years to mature and that is a lot of nibbling. Clothes moths don't fly very far, so they are more likely to have been introduced into your house on something than to have just dropped by.

COW PARSLEY OR QUEEN ANNE'S LACE

Alleys and paths can look beautiful at this time of year with a froth of white flowers on either side. This is cow parsley or Queen Anne's lace. Cow parsley refers to the leaves and simply means parsley for cows; the lace connection is obvious to anyone looking at the flowers, but no one is quite sure which Anne (or occasionally Ann) the common name refers to – probably Queen Anne who reigned 1702–14. When Queen Anne travelled the country in May, the people said the roads were decorated for her, and the pure white flowers may also be a sad reminder of her many pregnancies but no surviving children; whatever the reason, we think it is a far prettier name. Cow parsley is edible, but be careful as giant hogweed, which looks similar, is very poisonous.

CYCLE PATHS

As committed cyclists, we are much in favour of the network of cycle paths popping up across towns and cities. Swapping the car for a bike is doing your bit to improve air quality and nature will thank you for it. Travelling at a slower pace also gives you a chance to look around a bit more and really see what is out there. On open roads this can be hazardous, but while we wouldn't recommend completely ignoring other cyclists and pedestrians, the cycle path makes things safer and give you a good opportunity to enjoy your surroundings.

25 MAY

HAWTHORN BLOSSOM

This is the spring month of warmth as well as light, although the warning 'Ne'er cast a clout till May be out' is one to heed. *Clout* is Old English for 'clothing', and May almost certainly refers to hawthorn or May blossom, rather than the month as it will not flower until the weather is reasonably warm and settled and heading towards summer. You are more likely to see May blossom, or fairy thorn as it is also called, in its full, white, frothy glory in a town than along a country lane, as the plants bloom on second-year growth, stems which are frequently pruned to keep hedge plants in check.

26 MAY

EGYPTIAN GOOSE

Tony and Cleo, the Egyptian geese in our park, have produced
a brood of three ducklings. Egyptian geese are technically ducks
rather than geese, and these three look particularly cute scampering
along behind mum and dad. Egyptian geese, as the name suggests,
hail from the African Nile Valley but have taken to British life with
gusto and are now widespread. Their name references their origins,
but we like to think the chocolate patches around their eyes that
resemble the exotic eye make-up of Cleopatra, may also have played
its part. They are relatively sociable, hanging out in small flocks, and
although they are good swimmers they seem to prefer the land where
they feed on grass and leaves as well as worms.

A CUCKOO IN THE NEST

Cuckoos make the world's worst mothers. They never raise their own young, laying their eggs in the nests of other birds. Cuckoo eggs can be blue, brown, green or grey – this is because cuckoo eggs have evolved to mimic those of the host species, however, each female cuckoo can only lay eggs of one colour. When the cuckoo hatches, it will push any other eggs or chicks out of the nest (a very ugly stepsister indeed). Usually, cuckoos pick the nests of smaller birds such as dunnocks, pipets or robins as their host, but we once saw what we thought was a cuckoo egg in a Canada goose's nest – good luck little cuckoo with muscling out those siblings we thought!

ELDER TREE

Elder trees often look like large shrubs, growing untidily around park edges. Any scruffiness is forgiven for their double harvest – flowers in spring and berries in autumn. The small, white flowers grow in flattish clusters and, if you want them for elderflower cordial, you need to pick them just before they are fully opened.

The tree has biblical connections: the wood from an elder was used to make Jesus's cross and Judas hanged himself from an elder tree. Like many other trees, it is reputed to have connections with witches, and if you walk past an elder tree you should raise your hat and bid it good day. We tend to observe this tradition, as it seems only polite to greet a tree that provides the ingredients for both elderflower cordial and elderberry wine.

GRASSHOPPER

The song of the grasshopper is the sound of summer and, if we are lucky, now is when it will start. Grasshoppers are small insects, and you may struggle to see them in the long grass, but they make a lot of noise for their size. They are actually louder in towns and cities than in the country, trying to make themselves heard above the noise of traffic, which is a similar pitch to some of their songs. Their singing is the result of rubbing the little bumps on their legs against the veins of their wings to make a chirruping sound and the males do it as part of their courtship ritual. Those legs are powerful too. A grasshopper can catapult itself off its rear legs and jump up to 200 times its own height.

PEACOCK BUTTERFLY

The beautiful peacock butterfly is a species that has definitely benefited from rewilding in our parks and public areas. It lays its eggs on nettles and when the caterpillars emerge they build a communal web on a nettle patch to shelter in until they are large enough to survive alone. A caterpillar hanging from the underside of a nettle leaf like a tiny hook is almost certainly a peacock larva about to form a chrysalis.

Once they emerge fully formed in summer they are one of our loveliest and most popular butterflies, their rusty red-brown wings decorated with the peacock-feather-style eyes. They enjoy the sunshine for a short while before they hibernate for up to six or seven months.

RED VALERIAN

Confusingly, the flowers of this plant can be red, pink or white. Common valerian, a separate plant, is pink and not usually as common in urban areas. Red valerian was probably introduced in England in the 16th century from the Mediterranean, but escaped over the garden wall so long ago that it seems like a native. It grows in the cracks and crevices of rocky cliffs, but has adapted quite happily to take advantage of similar cracks in old buildings, loose stonework or garden walls. The leaves are a grey-green and the flowers grow in clusters on stems that wave merrily in the breeze.

JUNE

June can be beautiful, still early enough in the year for things to be fresh and green (sometimes rinsed by summer showers) and with longer daylight hours allowing plenty of time for after-work picnics, riverside drinks or long walks across the common. If you are really keen, you can get in touch with your pagan forbears and celebrate the solstice. There are lots of flowers in bloom, and with them come butterflies and other insects. Remember to sniff the freshly-mown grass as well as the roses.

———————

Rain falls, then it is gone.
The sun's bright on wet roofs, every washed tree
Has June's hard highlights, while a rivulet
Runs down the road that has been dry a month.

From 'Drought' by David Holbrook

1 JUNE
COMMA

While the comma, like so many punctuation marks, is losing ground in our social media communications, the comma is still one of our most common butterflies. It is easily identified as it is the orange-brown, raggedy one that looks as if something has taken a bite out of its wings. The white comma mark, which gives the butterfly its name, is less obvious as it is on the underside of its wings.

It is worthwhile looking under nettle leaves for comma caterpillars as these are particularly attractive. They are dark brown-black with orange and white markings that resemble a row of eyes and little white spines sticking out. Don't touch though, the spines are defensive and can irritate or cause an allergic reaction.

2 JUNE
FALSE BARLEY

False or wall barley is a grass that looks a little like an escapee from an agricultural field. It will grow beside paths, through cracks in pavements and in gaps in walls, setting seed in the autumn and growing afresh each year. The feathery flower spikes wave gently in the slightest breeze during June and July. They look soft, but are sticky to the touch and will attach themselves to any passing human or animal in a bid to disperse the seeds as widely as possible. False barley is used as animal fodder, but once it has flowered and set seed not even goats will eat it.

APHID

If you thought great-granny with her 12 children had it bad, save your compassion for the aphid. Aphids can give birth to up to eight babies a day and those babies can be born pregnant (don't even try to get your head round that one). They are expert kick-boxers, fighting back at predators by hitting out with their rear legs.

Aphids feed on the sap in plant stems by sucking it up with their mouthparts. As sap mainly comprises sugar, this makes the aphid very sweet, and it excretes a honeydew turning it into a tempting little sweetmeat for a wide variety of predators including ants, ladybirds and wasps.

LOOK DOWN AT DETAILS

Ideally pick a warm dry day and lie down outside. No, don't lie on your back and fall asleep, lie on your front, with a good view of some longish grass or undergrowth, a flower bed or a stretch of water and take a detailed look at the miniature world before you. Depending on your choice of location, there may be frenzied activity by beetles, bees and birds, or relative stillness with just the passing breeze moving the plants. Notice the details: the patterns on the plants; the different shades of green or brown; the shapes of the flowers; the movements of the creatures. This is a world that is hidden to anyone whose viewpoint is higher than a few centimetres off the ground.

5 JUNE

BRAMBLE FLOWERS

Brambles or blackberries belong to the rose family and, like so many other berries in that family, they are notoriously promiscuous. Blackberries, raspberries, tayberries and loganberries can easily be interbred by gardeners and, in the wild, blackberries will happily interbreed among themselves, to such an extent that several hundred microspecies exist in the UK. This means that each bramble thicket you see may be a mix of a number of slightly different plants. Later in the summer all anyone will care about are the berries, but for now look at the flowers. The long, arching, thorny stems bear pretty flowers along their length in either pure white or varying shades of pink. All being well, each flower will become a berry but, for the moment, just appreciate their beauty.

6 JUNE
TUFTED DUCK

There are some tufted ducks hanging out on the pond in the park. One of our favourite waterfowl, these small, diving ducks have a very distinctive head with a long, slicked-back pony-tail (the tuft that gives them their name). Males are black and white, sometimes with a purple-green sheen and a yellow eye, and females a more drab brown. They are huge show-offs, especially in the breeding season over the summer, when the males will dip their bills in and out of the water, bob their heads or, hilariously, glide past females with their necks stretched out, which is clearly screaming 'Look at me' in fluent duck.

7 JUNE
GORSE AND BROOM

If you are on a common between April and July and smell vanilla, look around for the yellow flowers of broom. Gorse looks similar, but flowers for most of the year and releases a delightful scent of coconut on warm days. Both thrive on the poor soils of common and heathland as they have the ability to take nitrogen from the air and convert it into nutrients. Broom's claim to fame is that as *planta genista* it gave its name to the Plantagenet dynasty of monarchs, while on a more mundane level its long stems were bound together for sweeping. Gorse is best known for the saying 'When gorse is out of bloom, kissing is out of fashion' – luckily this is the case for only a few days in the year.

BADGER

We would love to say we had spotted an urban badger with its Art
Deco-striped face, but sadly that treat is yet to come. Urban badgers
are out there, but not easy to find. There are even towns named after
them: the names of Brockenhurst and Brockhall are derived from
the old word for badger. Badgers are happy to build setts in parks,
gardens and golf courses. Come night-time they are out foraging for
snacks. They mainly eat worms (a full-grown badger can eat up to
200 a night), but city dwellers have adapted to a wider diet than their
country cousins, feasting on small rodents, snails, bulbs and even cat
food. The key to successful spotting is to stake out the sett about an
hour before dusk and be patient.

9 JUNE

ELM

Most national collections of plants are held in gardens or parklands, but unusually the British national collection of elms is in a city – Brighton. Elms like coastal air and the position of Brighton, cut off from the rest of the country by the South Downs, means that the trees have been less badly affected by Dutch elm disease, which arrived in Britain in the 1970s and has left us with a land of low, short-lived trees. Elms spread by suckering rather than producing seeds and are susceptible to mass disease. Up to the age of about 20 and a height of about 9 metres (30 feet) the trees remain healthy as the disease-spreading beetles tend to attack mature trees. The stumps regenerate and new trees grow up, some of which appear to be resistant, so there may yet be hope for the reappearance of tall, majestic elms in our parks.

10 JUNE

WATER LILY

There are white or yellow water lilies, both of which prefer still or slow-moving water. The glossy, heart-shaped leaves float on the water, as do the large, white flowers, which can reach 20 centimetres (8 inches) across. The yellow flowers are smaller (5–6 centimetres [2 inches]) and grow on stalks that raise them just above the water level. Interesting uses for water lilies in the past included steeping the rhizomes in tar to guard against baldness and eating the seeds to preserve chastity. Both of these were widely used during Elizabethan times, but we wouldn't recommend them now.

WEEPING WILLOW

Weeping willows resemble huge green hairy creatures lurking by rivers or in damp areas. They are garden plants rather than truly wild and there are a number of different cultivars. All have drooping branches, long thin leaves and long yellow catkins. They like their roots to be wet and require plenty of space to be seen at their best. Hunt them out in parks and riversides. They are perfect for dens or secluded picnics in the cavern-like interiors.

12 JUNE
JAY

The jay is nearly a lot of things, but the sum of its parts makes it very much its own bird. There is a hint of budgerigar in the bright blue and black striped edges to the wings; the soft pink-grey plumage recalls a parrot; and then its harsh, screeching call reminds you that it is actually part of the crow family. Most amazingly it hides away hundreds of acorns for winter – does this bird think it is a squirrel? The jay has yet another unusual behaviour: it loves to roost on ant's nests allowing the ants to crawl through its feathers, cleaning it of parasites. What we love most about nature is there is always something to surprise you if you look closely enough.

13 JUNE
FOXGLOVE

Wild foxgloves, which are nearly always purple, ideally like open woodland but are not fussy, and any rough ground will do. The plants are biennial, growing a base of leaves in the first year followed by a tall flower spike reaching up to 2 metres (6½ feet) the following year. Inside each tube-shaped flower on the spike, you can see intricate, spotted patterns, which attract bees and pollinating insects. The insect lands on the lower lip of the flower, walks up the tube, picking up and dropping off pollen as it goes. Foxgloves contain the chemical digitalis, which is used to treat heart failure and high blood pressure, but they are poisonous if eaten, giving rise to the saying that they can 'raise the dead and kill the living'.

14 JUNE
CARP

We kept goldfish as children, but they were only the size of a thumb. Hanging over the bridge we peer into the river and see two monster carp. These are no goldfish, although they are related. Originally introduced into Britain in the medieval period as a source of food, carp are now common in canals and waterways. They can live up to 50 years and weigh up to 23 kilograms (50 pounds) – as much as a fully grown male bulldog! Gliding gracefully through the water, the carp grazes on algae and water plants and nibbles small fish, which it senses using the thin, whisker-like barbels projecting from either side of its mouth.

15 JUNE
DOCKS AND PIERS

As a trading nation, Britain once welcomed ships from all over the world, bringing silks and spices. That port infrastructure has left an urban legacy in the shape of dockside apartments and wharf restaurants, which service both people and wildlife. They are perfect spots for an urban nature watch and you can usually get a pretty mean latte while you observe. Projects to clean up the waterways have resulted in the return of bream, pike, perch and even the occasional salmon. In turn the availability of fish is a boon to wildlife, and alongside the ducks, coots and gulls you can see cormorants, kingfishers or herons, and sometimes even seals.

16 JUNE
BRACKEN

Bracken is basically a fern, one of those plants that has spread so well it is loved and hated in equal measure. Most are 1–2 metres (3¼–6½ feet) tall with fronds that are deeply toothed and hairy underneath. Like most ferns, the young fronds form tightly furled fiddleheads that gradually uncurl as they grow. In autumn, some fronds remain green, while others turn a rusty red. Traditionally bracken had numerous uses: flooring, linings for baskets, bedding for cattle, plaited as the soles of shoes, shelter for open-air campers and even soap made from balls of the ashes. Now most of these have been superseded, although making bracken sandals can be a fun way to pass an afternoon on a common.

GEOFFREY GRIGSON, THE ENGLISHMAN'S FLORA

This is not a field guide as such, but it is one of our favourite flower books. First published in 1958, it looks at the flowers family by family – are they native and, if not, when and how were they introduced? What are their common names? What are the stories behind those names? As the dust jacket on our elderly copy says, 'it will be irresistible to the amateur of wild flowers, and to all whose attitude is aesthetic and romantic and not baldly scientific. It is delightful, surprising, and curious reading'.

Geoffrey Grigson was a poet and scholar, as well as a naturalist, and he was as interested in a plant's historical uses as in its correct identification. We now consider many flowers primarily for their beauty; in historical times most were valued for their significance in medicine, cosmetics, superstition and seasonal celebrations. What you picked, when you picked it and how you used it were all supremely important. This book contains fascinating information from Greek and Roman naturalists, the works of medieval monks, famous herbals, such as those by Gerard and Culpeper, and handbooks of folklore.

In this very personal book, he includes extracts of poetry and encourages the reader to look at paintings, sculptures and tapestries. In the introduction he says it was a book he wanted and, as no one else would write it, he wrote it himself. We are sure you will love it as much as we do.

It is out of print, but available in libraries and secondhand.

PEACOCK

Peacocks provide a taste of the exotic. You certainly won't find one on every street corner, but there are plenty in towns and cities: in parks, botanical gardens and stately homes. They are truly majestic and when the male displays his tail feathers with their blue-green eyes it really is something special. The birds are at their best from late spring to early summer during the breeding season when their plumage is at its most vibrant. A mature peacock can have up to 200 feathers in his tail, which will vibrate when he raises it for display (this is known as train rattling). A peahen not only sees the display, she also feels it in her crest, which acts as a sensor.

19 JUNE

GARDEN TO FERAL

Most of our garden plants originally grew in the wild; they have just been bred to suit horticultural needs: larger, brighter flowers, controlled size and the ability to repeat flower. Given half a chance, many will happily escape and revert to their wild state. The most obvious escapee is the buddleia or butterfly bush. Any crack in a building, any railway embankment or any deserted patch of ground will do. The plants are straggly because no one prunes them; they have brown seed heads because no one deadheads them, but they survive and show the flimsiness of our supposed control over nature. It is almost as if the plants are laughing at us.

20 JUNE
SUMMER SOLSTICE

Today (or tomorrow, depending on the relative positions of the earth and the sun) marks the point of the longest period of daylight and shortest night of the year. In some countries it marks the middle of summer, in others the start of the season, but since prehistory it has been observed with festivals or rituals. Many ancient peoples lit bonfires, hoping to strengthen the sun for the remainder of the growing season until harvest. Other gathered at stone circles or monuments to celebrate. You can watch the sun rise and wash your face in the dew – this will be at about 4.40am – or you can celebrate later in the day. Regardless of the weather the light will be there, so make the most of it.

21 JUNE
FRESHLY MOWN GRASS

For many people, the smell of freshly mown grass is the smell of summer. There are scientific explanations behind the scent, which is actually a defence reaction from the grass to repair the cut blades and prevent further damage, but for us it is a joyous smell. Early in the summer, before the grass becomes parched and dry, find a recently mown park or public garden and walk barefoot. Many of our ancestors may have been forced to go barefoot but for us it is a luxury, the soft blades of new grass gently tickling the soles of our feet.

22 JUNE
GOLDFINCH

A pretty bird with its red face and golden-yellow wing patches, the goldfinch has enjoyed some lovely names. The early Anglo-Saxons called them thistle-tweakers due to the long, slender beaks, which they use to prise difficult-to-get seeds out of the heads of thistles and teasels. Later in the Tudor period, the goldfinch was known as King Harry's bird or King Harry's red cap, from its brightly coloured, regal plumage. It has long been a symbol of good fortune (the collective noun for goldfinches is a charm), so if you fancy boosting your good luck try lacing your bird feeders with sunflower kernels or nigella seeds, which are their favourites.

23 JUNE
DOG ROSE FLOWERS

The dog rose doesn't seem to have much to do with dogs; the flowers that appear in early summer are white or pale pink and dainty, with five petals. It can grow a metre or so off the ground or scramble through trees reaching up to 9 metres (30 feet), clinging on with strong, curved prickles. Despite its scrambling and rampant habit, it is not the rose that enveloped Sleeping Beauty's castle – that was the briar rose or eglantine, which has a sweeter scent, its leaves smelling delightfully of apple after rain. However, you are probably fairly safe calling all wild roses dog roses as they are by far the most common. The scientific name, *Rosa canina*, possibly comes from the belief that the root of the plant could be used to cure the bite of a wild dog – *canina* is Latin for 'dog'.

MIDSUMMER'S DAY

Confusingly this is similar to, but not quite the same as, the summer solstice. It too has pagan roots marking the midpoint of the growing season, halfway between planting and harvest, but it is more about growing than the length of the day. Bonfires were lit and parades and parties held. It is also the birthday of John the Baptist and is one of the old Quarter Days when school terms started, servants were hired and rents were due. Whatever its roots, Midsummer's Day is an excuse to get outside and enjoy the long summer days, particularly if it was raining at the summer solstice or you were busy with other things. Celebrate outside with a picnic, a drink or a walk.

25 JUNE

LIME

The common lime or linden found in cities should not be confused with the small citrus trees that produce edible limes. Sadly, the latter would not survive in most urban areas in Britain outside a greenhouse. Even without fruit, the lime has much to recommend it: pretty, asymmetrical, heart-shaped leaves, white flowers that are pleasantly honey-scented and distinctive, round brown seeds in autumn. If your council is not too assiduous with its tree maintenance, you will be able to recognize most limes by the abundance of new stems around the base of the trunk. Common limes in streets rarely reach their full height of 50 metres (164 feet), but they still form imposing trees.

HOUSE MARTIN

House martins are the canaries of the modern city. They are responsive to air pollution, and left London in the mid-20th century when the smog was bad and have now returned, which must be a good sign for all of us. They are flighty little things and are constantly on the wing swooping after insects. They fly higher than swallows and can be distinguished by their white rumps.

The name comes from their predilection for building nests under the eaves of houses. The nests are made from hundreds of little mud pellets. House martins reuse the same nests year after year. A summer visitor, they arrive in April but will be gone by October, so enjoy them while they are here.

GREATER CELANDINE

The greater celandine may have been introduced to Britain by the Romans, or the plant could have been here before them. For thousands of years, it has been a popular garden plant, but to quote Geoffrey Grigson in *The Englishman's Flora,* 'Once in the garden it is now outside the fence.' A member of the poppy family, it is confusingly only distantly related to the lesser celandine, which is a cousin of buttercups. This celandine reaches about 90 centimetres (35 inches), can be found growing in cracks in pavements or walls, and has flowers resembling large buttercups. Break open the stem and you will find orange latex. Its Latin name *Chelidonium* comes from the Greek word for 'swallow', possibly because it flowers at the time the swallows arrive, or because ancient Greeks and Roman naturalists believed mother swallows used it to restore the sight of blind nestlings.

MONEY SPIDERS

Tiny little money spiders, with their shiny black bodies and brown legs, are reputed to bring good fortune so it is well worth making their acquaintance. They are also known as sheet weavers from the shape of their webs, which are often seen close to the ground, wrapped around garden plants or patches of long grass.

On late summer and early autumn evenings, you may well see money spiders 'ballooning' along on the breeze. The spider lets out a strand of silk that is taken by the wind and lifts the spider up into the air. If a travelling money spider lands in your hair, tradition says that to maximize your luck you should carefully spin it around your head and let it go.

CORNFLOWER

In the good old days, hay meadows were sprinkled with red and blue – poppies and cornflowers – which have now largely been banished from the sterile fields of our countryside. To be fair, cornflowers had been regarded as a pest since the 17th century, and even the poet John Clare had turned against them as an invasive weed, but they are tough plants and, given a chance in a garden or park, they will flourish. They are often included in wildflower seed mixes, and you will see their blue heads merrily bobbing among the longer grasses in the sunshine. The stems and undersides of the leaves are pale grey and felty to the touch, the flowers mid- to dark blue with pointed petals and darker or violet centres.

LITTLE RINGED PLOVER

Little ringed plovers are tremendously sweet birds that like to hang out in slightly less than salubrious spots. They are also something of a conservation success story, having colonized Britain only since 1938 and built up a healthy population around gravel pits and sewage treatment stations. Small, plump waders, they have black-and-white faces and brown backs with pale legs. It is the yellow ring around their eyes that gives them such a jaunty air and, along with the lack of a white wing bar, distinguishes them from other plovers.

Little ringed plovers hollow out shallow depressions in the ground, ideally in sandy shingle, to lay their eggs. Artificial reservoirs and gravel pits are ideal nesting sites. Both males and females will incubate the eggs. For such a small bird, little ringed plovers are aggressively territorial and will go to great lengths to protect their nests, luring predators away by feigning injury themselves.

JULY

There is pageantry to July, so get out and enjoy it. Forget green, this month the colour is red, from the coats of the swan uppers on the Thames to the clouds of fragile poppies round abandoned building sites and the tiny little scarlet armoured ladybirds. There are frogs singing by the ponds on summer evenings – no, don't kiss them, it is a myth and will not end well. The most spectacular thunder and lightning can occur now and make a note of the weather on the fifteenth, St Swithun's Day, and see if it really holds for the next month.

————

Today the queen ant and her lovers
took their nuptial flight, scattering
upwards like a handful of cracked
black peppercorns thrown in the face
of a bear, the bear in this case
a simile for the population of Lewisham
and Hither Green.

From 'Love Song, 31st July' by Richard Osmond

NATIONAL MEADOWS DAY

Traditional hay meadows have largely been replaced by 'weed-free grass mixes', which means that vast swathes of the countryside have lost their wonderful, diverse flowery fields. However, there is hope, as wildflower meadows are now being preserved and created in both cities and the countryside. A meadow can be a huge field or a tiny window box and at this time of year they will be at their best, brimming with soft grasses, dainty wild flowers and happy insects. Visit Plantlife (plantlife.org.uk) to find a meadow near you. This is a brilliant charity that protects wild flowers, plants and fungi, holding many meadowy events at this time of year.

2 JULY

MAGPIE

Even the laziest or most unlucky bird spotter in the biggest city can bag a magpie – they are everywhere and are very distinctive with their black-and-white plumage and long tail. However, take a closer look next time, and you will see purple-blue and green in their feathers. When spying a magpie, it is traditional to ward off ill luck by saying, 'Good morning, Mr Magpie, how are Mrs Magpie and all the other little magpies?' or similar. There are a lot of superstitions associated with this bird and a single magpie has been considered a portent of doom since time immemorial. This is supposedly because the magpie was the only bird not to sing to Jesus when he hung on the cross.

BUDDLEIA

These are the opportunists of the plant world. Look up at any unkempt building and you may well see a straggly shrub sprouting from a crack in the wall or roof: that is almost certainly buddleia. Railway embankments, building sites, disused land – nowhere is too rough. In gardens the flowers range from white, through every shade of blue, purple and even crimson and, if pruned, form neat bushes, but their wild cousins are leggy with predominantly purple flowers and brown seed heads. The flowers are long and pointed, looking a little like the delicate beak of an exotic bird and appear throughout the summer. Buddleia is known as the butterfly bush for good reason, you may well see red admirals, peacocks and small tortoiseshells flying round them enjoying the nectar.

GRASS WHISTLES

Blades of grass can be used to make what is probably the cheapest musical instrument. Find a flattish blade that is slightly longer than your thumb. Wedge it firmly between the tops of your thumbs and the knuckles at the base. Then create an air chamber by putting together the base of your hands and the tips of your fingers. Blow and, with luck, you will create a reedy sound that is neither bird nor human. Different grasses and different angles will create different sounds. While unlikely to earn you a place in a national orchestra, grass whistling is a skill we should not lose.

5 JULY
STAG BEETLE

Britain's largest beetle (don't quiver, we are only talking up to
7 centimetres [2¾ inches] here), the stag beetle does look slightly
intimidating; it is the male that has the reddish-brown, antler-like
mandibles. Interestingly, these jaws are used for fighting not eating.
For nourishment, they lick moisture from sap and rotting fruit.
In mythology, stag beetles have been associated with the Norse god
Thor, and legend has it if you put a stag beetle on your head you are
protected from being struck by lightning. However, you would need
to be a braver person than us to try it. During the mating season you
may be lucky enough to see male stag beetles wrestling to establish
dominance – they will tussle until the stronger of the two throws the
other to the ground.

6 JULY
CONVOLVULUS

In summer this plant bears beautiful, pure white, and occasionally
pink, trumpet-shaped flowers, but these are not enough to make
gardeners love it. A selection of this plant's common names says it
all: bindweed, hellweed, Devil's-guts – only in Wiltshire is it known
by the slightly more flattering name, granny's nightcap. As morning
glory (*Ipomoea* spp.) it is a pretty and controlled garden plant, but
wild convolvulus is on every gardener's hate list; even the tiniest
piece of root will quickly grow into a substantial and strangling plant.
Field bindweed has even prettier, pink-and-white striped flowers but
its roots go even deeper, so it is even more hated.

FRANCES HODGSON BURNETT, THE SECRET GARDEN

Forget the story of grumpy Mary and spoilt Colin, but instead read this as a tale of garden regeneration. Nature has taken over and created an enchanting wilderness in the walled garden but with a little gentle human intervention it becomes even more magical. The motto is clearly that we should work with nature rather than trying to impose our will on the natural world. Grow plants that will enjoy the conditions you have and, above all, avoid chemicals which will create an imbalance and make more work for you in the long run as you embark on a battle you cannot win.

MOSS

If you have a damp or shady lawn moss can be a pest but if you are creating a Japanese garden it is a vital symbol of harmony. In the wild it needs moisture but can survive in deserts, doesn't mind frost, absorbs pollution, uses its host for balance rather than nutrients as it lacks roots and inhabits a miniature world between air and earth. All of this makes it superbly adapted for multiple urban sites. Viewed close-up, (you will have to lie down) moss has the most beautiful miniature stems or flags that rise above the carpet of green to release thousands of spores into the air, some of which will eventually form new mosses.

COMMON BLUE BUTTERFLY

The common blue is, unsurprisingly, both frequently spotted and blue in colour. A small butterfly with really lovely bright blue wings fringed with white, and a soft velvety brown body, it is often seen on grassy commons or sports fields and patches of waste ground. It will be the blue colour that catches your eye, but look closely when it folds it wings as the underside is quite different, with black and orange spots ringed with white.

Usually, we are not people to dwell on scientific names, but the common blue or *Polyommatus icarus* is rather grandly named after Icarus, he of the Greek myth whose wax and feather wings melted when he flew too close to the sun. A heroic name for this little fellow.

BUTTERCUP

There are several different buttercups, but to quote an excuse used by many authors, their differences are 'beyond the scope of this book'. Just call them buttercups. The plants grow between 10 centimetres (4 inches) and 1 metre (3¼ feet) tall, and all have open, buttery yellow flowers. An easy way to tell whether someone likes butter is to hold a buttercup under their chin to see if it turns yellow although, in actual fact, this has more to do with whether the sun is shining than a person's particular taste. The flowers have a variety of local names, many of them butter-linked; traditionally buttercups grew in meadows, cows grazed meadows and so the link arose between the yellow of the butter and cream and the colour of the plant's petals and pollen.

THUNDER AND LIGHTNING

Summer storms are often more spectacular than winter ones. The sky seems to brood and thicken, then there is a flash of lightning followed by a rumble of thunder. The time between the two will tell you where the storm is – roughly five seconds for every mile away or three per kilometre. If the two happen simultaneously, the storm is right above you (and you are probably getting very wet). Thunderstorms are relatively common in summer when warm, moist air rises, causing massive instability and static electricity. Sheet lightning travels between clouds, fork from cloud to earth. A bolt of lightning is a little wider than a fountain pen and five times hotter than the sun. It takes the easiest path to earth which is why trees can be 'struck' and tall buildings have metal conductors.

12 JULY
CLIMBING TREES

Climbing trees is one of the things most grown-ups don't do; we grow
out of it. Which is a shame because it is one of the best ways of getting
close to nature and seeing the world from a different perspective.
Of course, climbing trees can be dangerous, harmful to the tree in
question, irresponsible and in some cases illegal, but if you check
your trees and are mindful of your abilities it can be a wondrous
experience. You don't even need to scale the heights, sometimes
a couple of metres off the ground is all you need. For further
information read *The Tree Climber's Guide* by Jack Cooke, which is useful,
charming, inspiring and witty; it's one of our favourite books.

13 JULY
POLLEN

We foresee the next revolution will be urban hay fever sufferers
uniting against town planners. Although the actual pollen count
tends to be lower in cities, the number of hay fever sufferers and
their level of suffering is proportionally higher. It is all the fault of
landscapers and their predilection for male trees. Female trees are
regarded as 'messier', as they shed more seeds, fruits or pods, but
it is the male trees that produce the allergenic pollen. If the gender
balance is maintained, all is well, but in cities all those lonely males
are releasing pollen into the air without female flowers to capture it,
resulting in an unnaturally high pollen count and lots of sneezing.
There is a suggestion that eating local honey can reduce your
susceptibility to hay fever, but as neither of us suffer from it we have
been unable to test this theory out.

DAMSELFLY

On a hot summer day, the electric blue damselflies (or bog dancers) hovering close to the river are beautiful. At least we think they are damselflies: it is difficult to tell while they are in motion. But when they settle, if they fold their wings they are indeed damselflies, dragonflies keep theirs extended. Damselflies are also more slender, although it might be rude to say so to a dragonfly.

Damselflies can be a variety of colours: red, brown, green, blue, black and yellow, and indeed the vibrancy of the colour changes according to light levels, but the electric blue ones are by far the most arresting. According to fossils, in prehistoric times damselflies could be as large as hawks, but today their wingspan stops at 15 centimetres (6 inches).

ST SWITHUN'S DAY

St Swithun's day, if thou dost rain,
For forty days it will remain;
St Swithun's day, if thou be fair,
For forty days 'twill rain no more.

In 862 St Swithun, Bishop of Winchester died and, at his request, was buried outside the church so he could feel the raindrops and feet of passers-by. In 971, on this day, the saint's day, his body was moved to a tomb inside. For the next 40 days a storm raged, and many people believed it was due to the saint being unhappy about his reinterment. Tradition now has it that whatever the weather on this day, it will be set for the following 40 days. There is no meteorological evidence to support this belief, but it is hard not to regard today's weather as particularly important.

16 JULY
POPPY

Each November, poppies remind us of the sacrifice made by so many in the world wars and many other conflicts, but it is in July that the wild flowers bloom. Their seeds can remain dormant for up to 40 years, waiting for a suitable moment. In the case of Flanders and many other European battlefields, it was the disturbance of the land through the action of war, which brought the seeds to the surface and, more sombrely, the blood and bones of the casualties that fertilized the soil. The flowers are a brilliant red with papery petals and can be seen wherever grass is left to grow long. The herbicides of intensive farming wipe out poppies so, like many countryside flowers, they have now found homes in urban parks and plots of disused land.

17 JULY
SWAN UPPING

Today we watched the scarlet shirts of the swan uppers as they headed out for this annual event that takes place on the River Thames. Upping is a tradition stretching back centuries and a vital conservation activity. The swan is a royal bird. The only other groups permitted to own swans being the Vintners and Dyers livery companies. During the Upping, the Royal Swan Marker (identified by the swan's feather in his hat) and his team go out in skiffs, when swans are spotted 'All up' is the cry, and the team surround and carefully lift out the cygnets, which they weigh, check for general health and injuries and then ring in an annual swan census. The ownership of cygnets depends on who their parents belong to, so families of swans have remained either Royal or Vintner or Dyer since the 12th century.

18 JULY

MUDLARKING

Such a glorious word for a fascinating pastime. Wandering along a river at low tide, eyes peeled to find what the river has uncovered is a private treasure hunt. The River Thames is rich in pickings but any tidal river wending through a city holds promise. Thames mudlarks date back to the 18th century when women and young children scoured the riverbed looking for coins or metal they could sell. Finding something unusual, attractive or even valuable is exciting but the real thrill is in forging a link with the history of the area. The riverbed has been keeping these finds safe sometimes for centuries and we love the fact that we might be the first person to see something since it was lost perhaps centuries before.

19 JULY

OTTER

Otters are timid, and unless you are in a wetland or wildlife centre, they can be difficult to spot, but their population is increasing and they are returning to rivers and canals, so you may be lucky. Look out for their five-toed footprints in muddy spots along waterways.

Otters are utterly endearing with their stubby noses, whiskers and powerful tails, and a group of them is known as a 'romp'. While adult otters are excellent swimmers they are not born to it. Young otters are often scared of water and must be taught to swim from around ten weeks old; once they get in though, they are naturals. They are able to close their ears and nose, which allows them to stay underwater for eight minutes, and they use their webbed feet to propel themselves through the water in an undulating motion.

FROG

During the winter, frogs hibernate beneath the mud of a pond or in piles of leaves or wood, but come summer they are out and about. Look around ponds in parks and gardens for this green-brown amphibian with bulging eyes. Those eyes mean they can see beside and very nearly behind them, they also have excellent night vision which should mean that pretty much nothing escapes an alert and hungry frog, particularly not a slow-moving slug. They use their long tongues coated with sticky saliva to catch their prey and roll it back into their mouths.

21 JULY

PERENNIAL WILDFLOWER MEADOWS

The word 'meadow' might suggest a romantic expanse of wild-flower-filled grassland ripe for haymaking, but urban meadows pop up where you least expect them, and are all the more delightful for it. You can encounter meadows between highways, around car parks, thin strips down the side of driveways or filling the moat at the Tower of London. Their mix of native grasses and wild flowers is attractive to pollinators and helps boost local diversity. Most wildflower meadows, regardless of size, are a mix of annuals and perennials. Annuals grow, flower, set seed and die in a single year, while most meadow perennials flower in summer, die down in winter and regrow the following spring, often living for many years. The perennials you are most likely to see are knapweed, oxeye daisy, clover, bird's-foot trefoil, buttercups, meadow cranesbill and cowslips.

THE BIG BUTTERFLY COUNT

Today feels like a good day for counting which is lucky as it is time for the Big Butterfly Count (bigbutterflycount.butterfly-conservation. org). This is a worthwhile thing to do as it helps conservationists get a more accurate picture of spread of butterflies in the United Kingdom. It will only take 15 minutes and you get a chart to help you identify different species.

There is an app (of course there is, there is an app for everything these days) but this is quite a handy one as you can count on the move. If you prefer, note down your findings on a piece of paper and enter on the website later.

Do remember this is a conservation activity not a competition. Don't hare off to the local butterfly house to guarantee you report more than your neighbour. Now is a good time to visit your patch and check out any butterfly action there or lack of it, go into the garden or just look out the window. Include moths as well as butterflies – if it flutters it counts.

RAGWORT

Similar to groundsel, ragwort has clusters of yellow, daisy-like flowers. The petals of ragwort are longer than groundsel, and it tends to be taller, but if the plants aren't growing next to one another that is of little help. Ragwort is very common in cities and there is both a London ragwort and an Oxford ragwort. In around 1700, the latter was brought to Britain from the slopes of Mount Etna in Sicily, and planted in the Botanic Garden in the city of Oxford. A hundred years later it was spotted growing wild on the walls *outside* the Botanic Garden. Like so many wild flowers it spread across the country with the development of the railway lines, which offered unintentional transport for the seeds and undisturbed growing sites on the embankments.

24 JULY

CIRRUS CLOUDS

Cirrus are the pretty, wispy clouds very high up making feathery patterns against the sky. One of our favourites, they are the whitest of white and look beautiful against a deep blue sky or taking on the colours of a sunset. Painters adore cirrus. In weather terms they indicate a change on the way, but although they are formed from ice crystals, they don't drop rain themselves. Cirrus clouds are known as 'mares' tails' and they do look a bit like horses' tails. An old adage states that 'Mares' tails and mackerel scales make lofty ships carry low sails', a nod to when sailors relied on clouds to warn of approaching bad weather.

WHITE BUTTERFLIES

The cabbage white or more correctly the large white, small white and green-veined white butterflies are often found around vegetable patches and allotments. Among our most common butterflies, they are attracted to all sorts of brassicas, not just cabbages. They have creamy white wings that look a little yellowish underneath, black or greyish wingtips and bodies and sometimes black dots on the forewings. The green-veined white, predictably, has green-grey veins on the undersides of its wings. They lay yellowish eggs on the underside of leaves (very bad news for your cabbages). If the weather is warm, you can get two generations of whites in a single year.

HOUSEFLY

We find houseflies simultaneously fascinating and repellent. Maybe it was watching David Cronenberg's *The Fly* back in the 1980s – that one had us hiding behind the sofa for days. The next time you see a fly, look closely, their eyes are huge. In fact, they are made up of thousands of individual lenses that combine to produce a single image meaning, we think, that a fly sees the world almost like looking through a kaleidoscope. Because of the size and complexity of their 'eyes' houseflies can actually see behind themselves and this, combined with super-fast reaction times, helps keep them safe from natural predators and the fly swat. And just to add to their strangeness, they taste through their feet, which may explain why they like to walk over their food.

27 JULY

ALGAE

Scientists have yet to come up with a universally accepted definition of algae; it is not a true plant and yet neither is it a fungus. In the sea the most obvious algae is seaweed, while on land the group includes a huge range of different organisms that live in water or on damp sites. Algae has an important role as the base of the food chain for aquatic organisms and it absorbs carbon dioxide and releases oxygen. So far so good, but algal bloom or thick layers can grow when excess nutrients build up in water, usually caused by human mismanagement. The bloom can cause the water temperature to rise, encouraging more growth in a vicious circle. It looks unsightly, and some of these algae contain toxins, which are harmful to humans, wildlife and the environment.

28 JULY

ROSEBAY WILLOWHERB

This is often found on the same sites as ragwort, with both plants happily making the most of rough, untended land. The tall, pink-purple spikes are actually made up of tiny flowers, each with four petals, the grey-green leaves resembling those of the willow. In autumn, clumps of the plant resemble bonfires as the leaves turn every shade of red and orange. It will happily colonize burnt land, earning it the name fireweed and during the Second World War its appearance in bomb craters earned it the name bombweed.

29 JULY
LADYBIRD

The old nursery rhyme 'Ladybird, ladybird fly away home' was a warning to this prettiest of tiny beetles when stubble was being burnt and their homes were in danger. The most recognizable ladybirds are the two and seven spot kind with their scarlet coats and black spots but if you look closely there are over 27 common species around in a variety of colours. Sometimes known as ladybugs, for many cultures they are considered harbingers of good luck.

The lady in the name ladybird refers to the Virgin Mary. According to some legends medieval farmers plagued by pests prayed to the Holy Mother and their crops were saved by little red bugs which they came to call lady beetles and later ladybirds. Others say the red of their backs represents her cloak and the seven spots her seven sorrows.

When stressed ladybirds excrete a yellowish oily substance from their knees (yes, knees, we thought that was weird too) which tastes smells and tastes unpleasant. It is also thought the pus-like substance might also be a kind of camouflage, making the predator think this ladybird is sickly and therefore far less appealing as a tasty snack.

ERMINE MOTH

Not all moths are equal and ermine moths are the black-and-white Dalmatians of the moth world. The caterpillars are creamy white with black markings and live between extensive webs in trees or on plant or leaf litter, but occasionally and most spectacularly, in some very odd places – we once saw an entire roundabout covered with moth webs. The adults, with their black dotted white wings, emerge in July and August when they mate and lay their eggs; the larvae overwinter under the protection of webs being fully grown and ready to start the cycle again the following summer. They are an outdoor moth that enjoys parks and gardens and, like most moths, are attracted to light so you might see adult females fluttering around your garden lighting.

31 JULY

LIZARD

You have to hand it to them, lizards are quick. One minute they are basking on a wall and the next, whoosh, a flick of the tail and he's gone. There is something sweetly primeval about a lizard, which looks like a little baby dragon. Reptilian sun worshippers, they are cold-blooded and cannot heat their own bodies, relying on the warmth of the sun to do it for them. Think of them as snakes with legs and like snakes they have scales. Common lizards are quiet little creatures, they have no voice and communicate with each other by touch: nudging, licking or biting. They are quite nervous and easily spooked. If caught by predators they can drop their tail, escape and grow another one. They grow young within their bodies and if you see a very fat lizard in June or July it is quite likely to be pregnant.

AUGUST

August in the city can be our favourite month. When everyone
else seems to decamp to the Med the streets become quieter, and
we become more aware of the non-human residents. Birdsong is
louder (or we can just hear it better) and you can listen out for
crickets. It might even be warm enough to swim in the natural
bathing ponds on Hampstead Heath or in the Isis at Oxford.
There is often a lazy, dusty heat haze over everything which slows
one down allowing time to think and write or draw. And that fruit
you meant to eat is surrounded by a cloud of whirling fruit flies –
move in and look closely at their tiny red eyes.

———————

The August sun streaming in,
so quiet you can hear the wagtails
hopping across the manicured lawns.

From 'Creative Waiting' by Roger McGough

EATING OUTSIDE

It's perfect picnic weather and time to get the hamper out. We love eating outside (even in winter sometimes – we hail from hardy stock) and take every opportunity for a park picnic, lunch in the pub garden or a table on the roof terrace. Fresh air sharpens the appetite – that's our excuse for tucking in anyway. One of the benefits of al fresco eating is the chance to fit in a bit of urban nature spotting between courses. Look around you: are those peregrine falcons swooping between the towers over there? Is that a sparrow or a dunnock? Is that a money spider abseiling down that wall? The loser in this game of nature I Spy buys the drinks.

SHAGGY INKCAP

Some shaggy inkcaps or lawyer's wig fungi are popping up in a stretch
of verge in late summer. As the name suggests, these have woolly,
bell-shaped caps sitting proud on a whitish stalk that resemble a
wig or bonnet. When they first come up the cap is a tight lollipop,
but as they grow this opens into a bell lined with gills that are white
or pinkish to begin with, but mature to a deep black. As the colour
deepens the gills liquefy and turn slimy (they will also do this within
hours of being picked, so don't). During the 17th and 18th centuries
this liquid was collected and used as ink. They occasionally grow
singularly but are more common in small clusters.

DADDY-LONG-LEGS

The daddy-long-legs is not technically a species, but rather a
generic name used around the world for several long-legged
insects. It is also a classic 1950s movie featuring Fred Astaire and
Leslie Caron, which you should watch if you get the chance. The
daddy-long-legs hanging around dusty corners of the house in
summer are crane flies, with small, oval bodies, transparent wings
and gangly legs (with knees!). Those legs are as delicate as they look
and do often break off. The legs don't regenerate but a daddy-long-
legs can live without one or two (they have six) but obviously there
comes a point...

KNAPWEED

The purple flowers of knapweed, both common and greater, give the plants the appearance of thistles without the thorns. They grow up to 50 and 80 centimetres (20 and 30 inches) tall respectively, and you will find them on verges, railway embankments or in patches of long grass, usually with a cluster of bees and butterflies enjoying their nectar. The rich nectar makes it a popular choice for wildflower meadows, but it can become a thug, pushing out the more fragile specimens.

SUNLIGHT

'Sunlight on my shoulders makes me happy', is more of a universal truth than a cheesy country song. There is something about sunlight that cannot help but make you feel better, possibly because when the sun hits your skin it triggers the release of serotonin in your brain, making you feel calm and improving your mood. The myriad of sun gods and goddesses and cults around the worship of the sun throughout history are testament to its importance in our lives. The feel of the sun on your body is energizing, but also consider shafts of sunlight in a dusty library, the dazzling glare of sunlight off the windows of tower blocks or the mix of sunlight and shadow along the pavement reflecting the movement of the clouds.

AVENUES OF TREES

Grand avenues of trees are traditionally associated with long drives leading to large country houses but there are a surprising number to be found in cities. They may be small avenues of delicate cherry trees or huge arches of beeches or planes towering over a suburban road. In the case of the latter, this is often seen when houses are built in the grounds of a much older property, and the original driveway becomes an ordinary road. Some streets give away their arboreal origins – Hawthorn Walk, Beech Avenue or even Cherry Tree Lane – and it can be interesting looking for the original specimens, although street trees are usually too widely spaced to form a proper avenue, and do not necessarily live as long as their names. Whatever the size or season an avenue of trees is impressive.

CUMULUS CLOUDS

Cumulus clouds are the ones you drew as a child – the cotton wool- or cauliflower-shaped ones suspended in a blue sky. They have flattish bases and are fairly low-lying. Despite the greyish tinge to the bottom of the cloud, they rarely bring rain and are known as fair-weather clouds. This doesn't stop them getting in the way of the sun but, as they move quickly, hold your nerve and they will soon pass over. Cumulus clouds can form long lines, up to 480 kilometres (300 miles) long, called cloud streets.

ORANGE TIP

The name orange tip is a bit of a misnomer: only the male butterfly has an orange tip to his wings; the female has a more dignified and restrained grey one. The main part of the wings on both genders is white with a characteristic black spot on the forewing, and an effective grey-green camouflage colouration on the underwing, making them difficult to spot when they are at rest.

Their pretty appearance masks the fact that they are cannibals. When the caterpillars emerge they eat their own eggshells, which is fair enough, but having got a taste for it will move on to eating any nearby unhatched eggs or small caterpillars.

9 AUGUST
GROUNDSEL

Unless they are growing side by side common groundsel and common ragwort are hard to tell apart. Both have small, yellow, daisy-like flowers that grow in clusters, and both are opportunists, keen to take root in any untended earth. As is the way with opportunists, you are likely to find groundsel growing dustily in cracks between paving stones or any waste ground. The name comes from the Old English *grundeswilige* or 'ground swallower' and, true to its name this plant will rapidly take over any bare earth. In the 19th century, before packaged pet food, weeds such as groundsel were collected and sold to wealthy Victorians for their caged birds.

10 AUGUST
HOVERFLY

Hoverflies are made for summer – that lazy, mid-air hover seems the perfect tempo for a warm afternoon in the park. At first glance they can look a little like bees or wasps, with yellow and black bands, but if you look closely you will see they have only one pair of wings (a characteristic of flies) rather than the two pairs found on wasps and bees. They also don't sting, so you can enjoy that hovering without fear. They are great pollinators and voracious eaters of aphids, so should be welcome in your garden. Hoverflies have their own Facebook page, UK Hoverflies, with over 6,000 members, so if you find yourself becoming a little bit obsessed you will be among friends.

11 AUGUST
CYGNET

It's easy to love fluffy little baby cygnets, but for sheer entertainment value you should revisit around August, when they reach their troublesome teens. Part-grown cygnets are gangly and awkward-looking with an amazingly flexible neck that they will twist round and rub over their backs. We stood for ages watching a pair of siblings in the dock basin hanging round a discarded traffic cone, ducking and diving, nipping at each other, showing off, attempting dad's stand-up wing flap and belly flopping back into the water looking embarrassed. By late autumn they may have gone, driven off by their fathers. Their brown feathers have turned white and, all grown up, they have the confidence to begin life on their own.

12 AUGUST
ANNUAL WILDFLOWER MEADOWS

Many urban meadows are planted using annual wildflower seeds, but in practice many of these areas will also include some perennials as well, as these plants will quickly colonize any suitable areas. The annuals you are most likely to see are poppies, cornflowers, corn cockles, corn marigold and corn chamomile. Yellow rattle is a particularly important plant as it helps to ensure the health of the meadow. These meadows should be cut after the flowers have set seed to ensure there will be new plants the following year. Keeping the balance of plants in a meadow can be tricky and many annual meadows are reseeded each spring.

13 AUGUST

METEOR SHOWERS

Meteors are nature's firework displays. Cosmic debris, some as small
as a grain of sand, heats up as it enters the earth's atmosphere and
what you see is the glowing trail of hot air. These spectacular events
do occur relatively regularly but of course you need to be in the right
place at the right time in the right weather conditions to be able to
see one. If you keep an eye on the Royal Observatory's Night Sky
Calendar (rmg.co.uk) you will be poised for an amazing celestial
experience. Ideally, take a flask to a local common just before dawn,
but anywhere with a good view of the sky and not too much light will
do. The annual Perseids shower, occurring every July/August when
the earth passes through the Swift Tuttle comet's orbit, is a good one
to watch out for.

14 AUGUST
FLYING ANT

It's Flying Ant Day. Triggered by heat and humidity, this is really more of a season than a day and can occur more than once in a year. Flying ants are not a separate species, they signal a specific phase in the life of most ants. And you don't need to keep your eyes peeled to spot them, when they are about it is like walking through an unmissable ant rave. Virgin queens leave the nests and meet up with swarms of males. The queen leads the males on an acrobatic chase (the nuptial flight) across country, mating on the wing with the bravest and best. When done she drops to the ground, chews off her own wings and crawls away to form a new colony. The males die, exhausted but happy.

15 AUGUST
NATURAL HISTORY MUSEUM

Throughout this book we have mostly avoided recommending specific places to go or things to do in particular cities. Our aim is to alert you to general things of interest rather than point you at individual places. The Natural History Museum in London is an exception. Not only is it an amazing museum, packed with incredible things, but it has a fabulous wildlife garden. Opened in 1995, it is surrounded by some of the busiest roads in the city, yet over 350 species of beetle, 550 species of moth and butterfly and over 150 species of fungi have been recorded there. Not to mention the countless birds and flowers. It is a delightful oasis of calm amid the roaring traffic.

16 AUGUST
HAIL

The energy in a hailstorm is astounding. A hailstorm is short and sharp, no more than five or ten minutes, but those balls of ice really dance, drumming on your roof and windows and bouncing up and down on the tops of cars. Oddly enough, the most common time for a hail storm is in summer, when warmer weather creates the energy to produce larger clouds. Hail is formed when frozen water droplets are driven up and down through tall cumulonimbus or storm clouds; as they rise they collect additional layers of ice until they become so heavy gravity causes them to fall. The next time you see reasonable-sized hailstones pick up a few and cut them open to see the layers of murky and clear ice like rings on a tree trunk.

17 AUGUST
KINGFISHER

Canals, ponds and sluggish rivers are the best places to spot a kingfisher. They love slow-moving or still water, ideally with overhanging trees, where they can indulge their passion for fishing. A kingfisher can eat its own body weight in fish in a single day. However, they are smaller than you think, often the size of a large sparrow and a flash of blue and orange when one is on the move is when you are most likely to see one. In folklore, they are associated with calm weather (their Greek name 'halcyon' comes from a princess from Greek mythology, Alcyone), and in the past kingfisher feathers were carried as talismans to ward off storms. Unusually for a bird, they nest in tunnels in the riverbank.

YELLOW RATTLE

Described unflatteringly in most wild flower guides as a semi-parasitic annual, this unassuming plant is vital to the success of most wild flower meadows. It weakens grasses, allowing prettier but less dominant meadow flowers to thrive. Without it many meadows would gradually become mainly grass and little else over the course of a few years. For this reason, many farmers consider it a pest, as it reduces hay yields, but in urban meadows it is invaluable.

It grows to about 50 centimetres (20 inches) tall and the yellow flowers are followed by a green-brown calyx in the shape of a purse with the seeds rattling inside. As the plant ages, the calyx dries and then opens with the seeds being launched into the wind.

19 AUGUST

MOSQUITO

On a muggy summer night, the drone of a mosquito can keep you awake worrying about the possibility of it landing and biting you as soon as you fall asleep (and bite it will, although unlike those in the tropics British mosquitoes do not carry infection, so itching is your only worry). There are over 30 species of mosquito in Britain, including a London Underground variety, which reputedly preyed on Londoners sheltering from the Blitz during the Second World War. The warmth of the Underground system suits the mosquito very well, but you also get them above ground in cities during the hot summer months. The mosquito has an internal body clock that regulates mealtimes, and they mostly prefer to feed at dawn and dusk when, attracted by your lights, they will fly into the house.

CUCKOO SPIT

Cuckoo spit is one of those truly wonderful examples of British nomenclature that makes us the nation we are. It is neither spit nor does it come from a cuckoo, but you will see it during summer if you look. Cuckoo spit is the white, frothy stuff that clings to the stems of various plants. It comes from the larvae of small, brown insects called froghoppers. They are called froghoppers because of their ability to make huge leaps if threatened. The cuckoo spit itself is the sap of the plant that has been ingested by the larvae, and then excreted to form a protective cocoon which prevents the larvae from drying out and protects it from predators. The spit itself will not harm the plant although the nibbling might.

RED ADMIRAL BUTTERFLY

The red admiral is probably the first butterfly you learn to recognize with its velvety, black wings and distinctive orange and white markings. They have a particular fondness for buddleia and other garden plants, and we often spot them at this time of year feeding or basking in the sunshine. When the sun goes behind a cloud, they close up their wings to preserve warmth. Like most butterflies, they only holiday in Britain, arriving from southern Europe and North Africa in the spring. The caterpillars are harder to spot, but look closely on nettles around May, and you may see some: black and spiny with a fine, yellow stripe down each side. It turns out there was no interesting admiral – the name is a corruption of what the insect was known as during the 18th century, 'red admirable'.

SHARING YOUR BERRIES WITH NATURE

Now is the time to start picking blackberries and timing is a delicate balance; you want the berries to be ripe, but you don't want other people to pick the best first. In your attempts to secure the best fruit, remember the birds and animals who depend on the berries for their lives, not merely as additions for pies and crumbles. Always leave plenty of berries on the plant for them.

We have found that Friday evenings are a good time to combine picking and a picnic supper as you beat the weekend visitors. We are obviously not revealing our favourite foraging locations.

NATURE RESERVES AND WETLANDS

Historically, most urban settlements grew up near rivers and close to most of these are areas of marshland. Add in reservoirs and gravel pits, and there are a number of wetland habitats close to towns and cities that are perfect for wildlife. Many are now protected, and some have been converted into reserves and wildlife centres. Unlike zoos, these are based on the natural conditions of the area, and attract local flora and fauna as well as the wildlife on show. At our local wetlands in Barnes, the resident heron knows when the otters are being fed and drops in for an easy meal. The sites also make a good stopping-off point for migrating birds. They may contain exotic species, but these reserves are also brilliant places to see and discover more about our native wildlife.

GULL

The strident call of gulls always makes us think of the seaside, but gulls (never call them seagulls) are becoming increasingly common in urban environments. For them, a city is just a series of rooftop islands with few natural predators, and bins and landfill sites offering a tasty buffet. Long-lived gulls often survive for decades, and experience combined with a keen intelligence allows them to learn and adapt easily to the challenges of city life. There are several different gulls, but in towns you are most likely to see the pale grey-and-white herring gull or their aggressive cousins the larger black-backed gulls (these are the ones most likely to swoop on your poorly-guarded chips).

FRUIT FLY

Something in the fruit bowl may be a little past its best as a small swarm of tiny flies has risen from over there. They don't actually eat the fruit, but rather the fungus that grows on rotting produce. Fruit flies are really tiny, but if you look closely, you can see that they have huge red eyes.

Despite their size, fruit flies have an important role to play in research. Their rapid life cycle and the fact that they share up to 75 per cent of their gene profile with humans, means they have played a significant role in genetic, disease, heredity and microbial research. At least five Nobel prizes have been awarded to scientists who based their research on fruit flies.

26 AUGUST
HAWKBIT

A lot of the plants that look like dandelions could well be hawkbits, hawkweeds or hawkbeards, all of which have the same bright yellow flowers. Rough hawkbits are hairy while autumnal hawkbits have smooth leaves and branching stems. They also continue flowering well into late autumn, long after many wild flowers have retired for the year. More unusual are orange hawkweeds, which helpfully are orange and therefore easier to identify. The takeaways here are that in the Middle Ages it was thought that hawks ate the flowers to improve their eyesight and the warning that a dandelion is not always a dandelion.

27 AUGUST
STRATUS CLOUDS

Stratus clouds are the low, grey blankets of cloud that hang around for days and, unless you love this kind of weather, tend to provoke a feeling of gloom. They hide the sun and although rain is not guaranteed is it not uncommon either. Rain from stratus clouds is not a short, sharp shower, it is that drizzly rain that seems to go on for ever. On the plus side, as stratus are low-lying the rain is not as cold as that from cumulus clouds and there is no hail, but if you are out in it that is small comfort. If you live or work high up, stratus are the blankets of cloud you can see beneath you, grey and gloomy compared to the blue sky up above.

SHIELD BUG

Shield bugs are seasonal camouflage artists. During the summer when you can find them basking in the sunshine they are green with brown wing tips (some species have red or brown markings or interesting stippling on their back) but, come autumn, they change with the autumn leaves to a bronzy colour. It is the flat, triangular shape of their armour-like backs that gives rise to their name. They feed on plants and other invertebrates, sucking out the juices in a vampiric manner. If disturbed, shield bugs produce a rather pungent-smelling liquid from glands in their abdomens, resulting in their less appealing name of 'stink bug'.

MULBERRY

These are park and garden trees rather than truly wild, but they are worth hunting out for their amazing berries. These look a little like blackberries but are soft, juicy, easily squashed and rarely found in shops or markets. Many mulberries were planted in London and other cities in the 17th century in an attempt to set up a silk industry but black mulberries were planted by mistake, rather than the white ones that are needed to make silk – unintentionally providing the locals with a great harvest. The trees quickly become gnarled and knobbly, often looking much older than their true age and needing supports to prop up the branches. A final word of warning – wear old or purple-coloured clothes when picking – the juice goes everywhere.

GOLDCREST

Goldcrests are tiny, well-camouflaged and flit about constantly, but if you are sensing movement in conifers then like as not there are goldcrests about. The goldcrest vies with the wren as Britain's smallest bird and wins by a whisker. Being small helps keep them out of sight of predators, as does their olive green and brown colouring, which blends into the foliage. But if you see that fluttering, look closely because they are really a very pretty bird. They have a golden yellow crest, which the male raises during mating displays, and a delightful dark moustache running down from their bills that somehow makes them look perpetually a little sad. Their cup-shaped nests are made from spider's webs, lichen and moss and look like charming hanging baskets for fairy folk.

SEEING IN COLOUR

Without nature urban areas would be an unforgiving grey but not all wildlife sees the world the way we do. Most mammals see shades of blue and yellow, with reds appearing as shades of grey, but their night vision is usually far better than ours, fine-tuned for hunting and avoiding being hunted. Rats keep one eye focused upwards as that is the direction danger is most likely to come from. Bees and butterflies can also see ultraviolet colours, important as many flowers have ultra violet patterns on their petals. The same is true for birds, who often see colours we can't. Finally, pity garden snails who can't focus or see colour – other snails and predators simply appear as a blur.

SEPTEMBER

September in the countryside welcomes the advent of the harvest season with bounty and fruitfulness. In town we may be grumpier as we have to return to work and the summer tan fades but there is still much to enjoy. Watch out for open days at local allotments where you can often pick up amazing produce and spend a happy afternoon making jams and pickles. Go to the park and collect conkers; you can introduce a young neighbour to the joys of play or just pop a few in dusty corners to deter spiders and moths. Celebrate rainbows or drizzle and pretend you are so over sunshine anyway.

———————

Even the rainbow has a body
made of drizzling rain
and it is an architecture of glistening atoms

From 'The Rainbow' by D. H. Lawrence

BOTANICAL PAINTING

A visit to the Marianne North House at Kew Gardens is both a treat and a source of inspiration. Marianne North was one of those intrepid Victorian women who defied convention and travelled the world painting people, places and, most importantly, plants. Most botanical painters work in watercolour, but Marianne used oils and her paintings have a vibrancy and impact that makes them special. She put the plants in context, capturing place as well as plant. Botanical painting is a lovely way to spend some time looking really closely at a plant and trying to capture its essence. If you are not keen on doing it yourself consider looking in antiquarian book shops, they often sell botanical bookplates and prints that are beautiful and look lovely framed.

2 SEPTEMBER

NIGHT

It's dark at night, less dark in town than in the countryside, but dark nevertheless. That certainly doesn't mean there is nothing to see. Nocturnal birds and animals are out and about, hunting under cover of darkness, and there are stars and planets to watch and the lovely velvety dark in itself is worth enjoying. Take some time to allow your eyes to acclimatize and if you can, avoid a torch, which will dazzle your eyes and frighten off darkness dwellers. Use your other senses too, your ears and nose – our sense of smell is heightened in early evening. Be careful though and keep to known safe places, sadly it is not only nature that is active at night.

NOT JUST A PRETTY FACE

Humans tend to be overly concerned with appearance: a spectacular
view, a beautiful garden, a stunning location. Nature couldn't care
less. Appearances in the natural world are for a reason: to attract a
mate, entice pollinators or ward off predators. The same applies to
their habitats; if a place has the necessary light, water or protection
it doesn't matter whether home is a manicured garden or a derelict
factory. Look more carefully at sites you would normally write off
as 'hideous'. Buddleia will grow in cracks in any building, foxes will
settle in any shelter and birds will happily nest in scruffy bushes
growing in waste land. For the natural world there is no division
between 'derelict' and 'wild'.

4 SEPTEMBER

ROSEMARY BEETLE

Not a friend of Colin the Caterpillar but rather a beetle that lives
on aromatic woody herbs, such as rosemary, sage, lavender and
thyme. What is most striking about this beetle is its iridescent
metallic purple and green stripes. It truly looks like a tiny, jewelled
brooch. If you see some this is definitely a photo moment. Sadly,
for the rosemary bush, these have become increasingly common in
Britain since the 1990s. They feed on both flowers and foliage in
late summer and early autumn and although they don't always kill the
plant they can do quite a bit of damage.

ALLOTMENTS

If you are thinking about getting an allotment, September is one of
the best times, even though you will probably end up in a queue for
many months after you apply. Even so, it is a great time to visit them
on open days. Blackberries, tomatoes, sunflowers and squash make
the plots colourful and jolly. It is very easy to be seduced by the idea
of having an allotment but be warned, battling brambles on a cold
winter's afternoon is likely to be your introduction to a new plot.
That said, it is a way of life that can become delightfully addictive
and, in return for your labour, will provide you with wonderful
fruit, veg and flowers for a fraction of the usual cost.

6 SEPTEMBER

BLOOD RAIN

The radio is forecasting blood rain, which sounds ominous, even
apocalyptic, and if we were medieval peasants we would be fearful
of the deaths of kings, but today it just means cleaning dust off the
car windscreen. A blood rain occurs when high concentrations of
reddish dust whipped up by strong winds is caught inside rain clouds
and then falls as rain, sometimes thousands of kilometres away. A
real blood rain, with concentrations of dust high enough to make the
rain look red as it falls and stain clothes, is relatively rare although
there have been documented cases. We get blood rains a few times a
year and on low news days the radio will talk them up. What you can
do though is collect your own little bit of the Sahara after the rain
has passed.

BLACKBERRY

It's blackberry season and time to return to those spots you scoped out earlier in the year, when they were just annoying brambles, to reap autumn's bounty. Be mindful when you pick and don't trespass or frequent spots that may have been sprayed. Remember to leave plenty for the birds and other wildlife who rely on foraging to live. Commons, towpaths and alleys are good hunting grounds. Gloves are useful as blackberry stems can cause nasty scratches. The berries are at their sweetest when they are a deep black.

Technically, and we do love a 'technically', blackberries are not berries but drupes. Each individual globe on a blackberry contains a seed, making it a separate berry and the whole thing a drupe. Whatever it is called it is delicious with cream.

TREEHOUSES

Treehouses are a luxury of childhood, nearly always belonging to other people, particularly characters in books. With a little imagination you can make a temporary house by a tree and enjoy some magical moments. Find a tree with wide, low, spreading branches, drape a rug or tarpaulin over them, spread a rug beneath and, hey presto, you have a house. It won't have the views of an elevated house, but instead notice the details of the trunk, the boughs above you and lie down and look at the world at ground level. Lunch, tea or evening drinks are obviously welcome additions. For inspiration read Italo Calvino's *The Baron in the Trees*, a magical fable about living in trees.

CLOVER

Red, white, crimson, hare's-foot, zig-zag, hop or shamrock – there are many different clovers, but all have three leaves, four if you are lucky – 56 leaves is the world record, found in Japan in 2009. Four-leaved or more are rare but do exist; if you find one look carefully as they often grow in clumps. The purple, pink or white flowers are ball-shaped, and if you pull the petals off and suck the base you will get a short but intense burst of honey, which is why the bees love clover so much. Looking slightly different, hare's-foot clover is fluffy and, unsurprisingly, shaped like a hare's foot.

CRAB APPLE

Crab apple trees are often found in parks, urban gardens and along residential streets. They form smallish, slightly scruffy trees, but in spring the blossom is gorgeous, the white flowers having pink undersides that give the trees a delightful, pinkish glow. The fruit ripens in autumn and looks like a small apple, but you will get a shock if you try to eat one directly from the tree – they are distinctly sharp. Once cooked, they can be added to pies and crumbles or, perhaps best, made into delicious jelly, which has a distinctive rosy colour. If you see crab apples in someone's garden, try knocking on the door and asking if they are using them. Often, people don't make their own jelly and they will be happy for you to take the fruit away, rather than have it rot under their tree.

MOON

The moon is a symbol of mutability but in truth it changes less than you might think. Because the moon rotates on its axis at the same rate as it orbits the earth, a phenomenon known as synchronous rotation, you will always see the same side of the moon. The other side really could really be made of green cheese.

Moonwatching is a fascinating pastime. In its 29-day cycle around the earth, the moon waxes and wanes through new to crescent, gibbous, full and back. You can watch the moon with the naked eye, examine more texture through binoculars or get up close to the surface with a telescope. And although you may think the moon comes in different colours, it really is just grey and white. The orange or reddish moons you see sometimes are caused by pollution, dust and smoke from our atmosphere.

SYCAMORE

These are large, majestic trees that you find in parks. Their
branches grow in clumps and, in summer, the trees can resemble
giant broccoli florets. In late spring the yellow flowers hang down
in clusters, appearing at the same time as the leaves, but it is the
winged seeds, which appear now, that are distinctive. They grow in
pairs at a right angle – a pair of straight winged seeds is more likely
to be from a Norway maple, the sycamore's close cousin. Sycamores
are tough and well-adapted to urban life, seeding and spreading via
suckers. They are beautiful trees, but they have one fault – the fallen
leaves tend to become a slippery mush. According to Network Rail,
sycamore trees are one of the six culprits that cause the dreaded
'leaves on the line'.

13 SEPTEMBER

TWILIGHT

Look up times for day and night and you will find a confusing array
of twilights in between: astronomical, nautical and civil. Each is
dependent on the exact height of the sun in the sky, but together they
mark the transformation from light to dark and vice versa. Built-
up areas rarely experience the true dark of night with an artificially
created twilight extending its time. Although the expression 'twilight
zone' is used to describe an urban area in a state of dilapidation or
economic decline, twilight is an attractive time of day in a city. The
colours of the city are muted, silhouettes stand out sharply against
the darkening (or lightening) sky, and it is a particularly good time to
admire the outlines of trees.

ELDER BERRIES

Having given us elderflower cordial in summer, elder trees now provide a second harvest of berries. The Romans used them as hair dye and the 16th-century herbalist John Gerard recommended them for weight loss, but we like them in jelly, vinegar, wine or cordial. The easiest way to collect the berries is to cut the clusters with scissors. Raw berries are mildly toxic, so don't be tempted to nibble while you harvest. These stems are poisonous so you should remove the berries; stripping them with a fork is the easiest way. You then have to pick out any unripe, green berries or wrinkled, over-ripe ones. All this palaver is worthwhile though, for the delicious flavour once the berries are cooked.

15 SEPTEMBER

CORMORANT

If you thought cormorants were only seen in Asian art, think again. We regularly see one on the Thames near Fulham drying his wings in the morning sunshine. Perched like this on a branch or pier, their large black wings outstretched, cormorants are very distinctive.

Cormorants feed by diving for fish, which they catch in their long, hooked bills, but their feathers are not waterproof like many waterfowl, and they need to dry off after a dip. They are expert divers and can go down to 46 metres (150 feet) shooting through the water using their wings as rudders.

16 SEPTEMBER

DEER

The royal parks such as Richmond and Bushy are home to herds of
red and fallow deer that have grazed there for generations, a legacy
of the parks' past as royal hunting grounds. They seem tame but
are wild animals and should be approached with caution. Deer are
not uncommon in woods and forests on the outskirts of towns and
cities and will occasionally wander closer in with reported sightings
in gardens.

Red deer are Britain's largest mammal, and a full-grown
stag is an imposing and majestic sight. Walking through one of the
royal parks on a foggy autumn day and hearing the bellowing of a stag
is eerie. Roe deer are smaller, with little, three-pronged antlers.

17 SEPTEMBER

MIZZLE AND DRIZZLE

It is one of those days, the ones that are wet but the rain is light, and
you need to go out but you don't care. It is drizzling or mizzling or if
you are in Scotland there is smirr. Even the words indicate it will be
fun. This is that kind of light mist-cum-rain that you *should* go out in,
as for the full experience you need to feel it on your skin or on your
tongue – go on, give it a try. Technically speaking, drizzle is where
the droplets are less than 0.5 millimetres ($^1/_{50}$ inch), but no one is
expecting you to get your tape measure out. Drizzle is more likely to
occur when the clouds are very low.

SEED HEADS

To encourage a succession of flowers on annuals and many perennials and roses, you should deadhead regularly, cutting off spent flowers so the plant produces more. Towards the end of the flowering season, it is worth leaving the fading flowers and allowing seed heads to develop. They will provide structure in your containers or flower beds through winter and can be a valuable source of food for wildlife. If you are lucky some seeds may even grow into plants for following years. Agapanthus, rudbeckia, honesty, sea holly and sedum all look wonderful as they fade. When the plants get tatty and start to collapse, simply cut them down near to the ground.

CRACKS IN THE PAVEMENTS

The trait that most successful urban nature has in common is that it is opportunistic. Foxes don't wait for a rental agreement to be signed before they move in, birds build their nests where they choose, and seeds will take root wherever they find a suitable spot. The next time you walk to your local shop, look at the cracks in the pavement and you will, unless the council has just blitzed the area, find a surprising number of plants growing there. They may be a bit straggly and dusty, but they have made the most of an opportunity that was offered to them – a tiny chink of light and just enough soil to take root. Count the number of different plants you see; identification can come later. For the moment just be amazed at the variety of life in a 'sterile' pavement.

20 SEPTEMBER

SILVERFISH

You may never see a shoal of silverfish (they are not really fish), but if you spot one you are looking at a link to our prehistoric past. Wingless insects, silverfish are one of the oldest creatures on the planet, predating dinosaurs.

A true city slicker, silverfish have adapted very well to urban life, inhabiting warm, damp corners in kitchens and bathrooms, but as they feed on starch or cellulose and will nibble at books and wallpaper, they are often an unwelcome house guest.

Their name is derived from their wriggly movements, which makes it look as though they are swimming, although in reality they run on six legs. They are less than 1¼ centimetres (½ inch) long but very fast. Lacking wings to fly, they developed this impressive turn of speed to escape predators.

21 SEPTEMBER

CONKERS

We are obviously going to ignore the health-and-safety aspect
of playing conkers; it's too much fun and realistically the risk of
injury is fairly low. Vinegar, baking and varnish will all harden your
conkers, but that is hardly the point. Thread string through them
and bash away. The old wives' tale of using conkers or horse chestnut
seeds as a moth deterrent is absolutely true. As the conkers dry out,
they act as a mild insecticide and are a good alternative to moth balls.
It is thought that the Vikings made soap from crushed conkers and
the Victorians used them for flour. We don't recommend this last
idea as conkers are mildly poisonous.

22 SEPTEMBER

AUTUMN EQUINOX

This is the time in autumn when the hours of light and dark are
equal. Much of nature will soon start to hunker down for winter but
there is still plenty to see. Look out for beautiful seed heads on plants
and autumnal berries, both ornamental and edible. On warm days
you may notice a flurry of activity as animals like squirrels lay down
extra stores for the colder, leaner months ahead and birds prepare to
fly south to warmer climes.

23 SEPTEMBER

PIGEON

Depending on your perspective, pigeons are the first opportunity for urban bird watching or rats with tails. In towns and cities, you rarely see a building or statue without a roosting pigeon, and they can be found in the most unlikely places: waiting for a train on a station platform or brazenly conducting a mating display in a supermarket car park.

Pigeons were one of the earliest domesticated birds and have been used for food, fertilizer, sport and message carrying. Thirty-two pigeons have been awarded the Dicken medal, which recognizes acts of gallantry or devotion in wartime.

Even experts can find it difficult to differentiate between male and female pigeons. In general, the males have a slightly more rounded head and are heavier, they also coo more.

RAINBOW

There is excitement when a rainbow is spotted: a colourful arc above the roofline, its bands of red, orange, yellow, green, blue, indigo and violet merging as though painted in with a watercolour brush. Our conscious brain knows this is a meteorological phenomenon caused by the refraction of rain droplets through light, but there is still a rush of childlike delight, and the feeling that if we could just find the end we would find that mythical pot of gold. Did you know you can only see a rainbow when the sun is behind you? Although it does need to be raining for you to see one, fortunately the rain does not need to be where you are standing.

25 SEPTEMBER

FIR CONES

Fir cones are the mature female flowers of conifer trees such as pine and spruce as well as fir. At the right moment they open, releasing the seeds which will, all being well, eventually grow into tall, majestic trees. Many cones grow over two years, with green buds appearing in the first year and mature cones the second. They vary in size and shape, and although some cones and their seeds are edible, you do need to be careful. Like foraging for fungi, it pays to seek the advice of an expert. We tend to use cones as ornaments, waiting until they are fully opened and have fallen to the ground, releasing the all-important seeds before we take them home.

SEAL

You are unlikely to encounter a seal on your way to buy a bottle of milk (we wish), but you can occasionally spot them in coastal or river cities. There are two main types in Britain, the common and the grey seal. Actually, the common seal is less common than the grey, and both can be grey in colour, so the names don't help at all.

Seal pups with their big eyes are adorable. Seals are inherently coastal, but declining fish stocks are forcing them to have to travel increasingly long distances upriver looking for food, and any you encounter are more likely in town on a shopping trip than local residents.

27 SEPTEMBER

BUTTERFLY HOUSES

Although butterflies will flutter around you in the park or garden, for a true butterfly immersion experience you will need to visit a butterfly house. There are several around the country, often in parks, zoos or botanic gardens, and these tropical paradises are well worth a visit. Protected from predators and harsh weather, butterfly houses can contain rarer and more colourful species, and give you the opportunity to see butterflies at the different stages of development.

If you want to help support butterflies at home, then think about growing nectar-rich plants in your garden or window box. Buddleia, Michaelmas daisy, oregano or French marigolds are all butterfly magnets.

FORAGING

Over the centuries foraging has evolved from a necessary life skill to a gourmet pastime. Even in the city edible fungi, herbs, vegetables and seaweeds are all available for free to an experienced forager. The key here is the word 'experienced'. It is absolutely vital that you know what you are doing and foraging should only be undertaken with someone who can confidently separate the delicious from the deadly. There are lots of classes and organised trips and time spent on one is fascinating and fun. There are rules: only take what you can eat, harvest carefully making sure not to damage any surroundings and respect private property. Also do consider the wildlife; you may be foraging for fun but make sure to leave enough for those who rely on nature's bounty for survival.

Most people think berries or mushrooms when they contemplate foraging but if you read up or go on a course you will find all sorts of surprising things that others may overlook. Wild chamomile, which looks a little like a green daisy that has lost its petals, has a surprisingly strong taste of pineapple, accounting for its common name of pineappleweed. It thrives in poor compacted soil on waste ground and is delicious in salads. Dandelions have a fresh peppery flavour and are common pretty much everywhere, gorse flowers make amazing tea (but do watch out for the spiky leaves which can be vicious) and rose hip syrup will revolutionize your ice cream sundae.

29 SEPTEMBER
DEADNETTLE

Rather than 'dead', these nettles should be described as 'friendly', as they do not sting. Red dead-nettles don't even look that much like true nettles as their leaves are less pointy and lack the stinging hairs. The leaf tips are often crimson, and from March until October the stems are covered with pretty, pinkish-purple flowers, the base of which will give a hint of sweetness when sucked. White dead-nettles look more like their stinging cousins, with flowers between the pointed leaves, but the leaves lack the stinging hairs and the flowers may also be harmlessly sucked for sweetness.

30 SEPTEMBER
PARAKEET

If you spot a flash of bright green darting about in the treetops it could well be one of the cheeky parakeets that have colonized our parks and gardens. They are native to Asia and Sub-Saharan Africa, but they seem very at home in the UK where the climate suits them. There is a charming urban myth that British parakeets are descended from birds that flew from the set of *The African Queen* in 1951, but it is more likely to be repeated accidental releases and escapes from aviaries. This is a pity, as it would be lovely to imagine that our parakeet's grandfather might have known Humphrey Bogart. They are adventurous eaters and the variety of exotic plant species found in gardens and parks makes them confirmed city dwellers.

OCTOBER

September can still feel like summer, but October is definitely autumn. Birds such as swallows may be departing for the warmth of Africa but others – geese, swans, ducks and starlings – all come to Britain for our comparatively balmy winters. There are berries to eat and admire, dew-drenched spiders' webs to marvel at, sweet chestnuts to roast and orchard festivals to enjoy. As the trees prepare for the cold weather ahead, this can be the most colourful time of year with the leaves' final moments of glory in every shade of red, orange and gold.

———

The trees are undressing, and fling in many places –
On the gray road, the roof, the window-sill –
Their radiant robes and ribbons and yellow laces;

From 'The Last Week in October' by Thomas Hardy

1 OCTOBER
HARVEST TIME

Gone are the days when we had a huge selection of apples, pears and plums to choose from; even most markets now only have a few varieties on sale from the over 2,000 grown in the UK. Events such as the London Orchard Festival aim to change this. Throughout the country at this time of year you will find orchard festivals; many individual fruits have their own special days, but the festivals now celebrate orchards in their entirety. If you choose your event carefully there can be fruit tastings, delicious food and drink, fun activities for all the family, a wealth of information and advice on getting involved.

2 OCTOBER
CARDINAL SPIDER

Cardinal spiders are the largest spiders in Britain and can grow up to 14 centimetres (5½ inches) – that is the size of your hand. However, size is not everything, and these are definitely less scary than they look. Like most spiders they can (but rarely do) bite and if they do the bite is relatively painless. The name reputedly comes from the legend that this spider terrified Cardinal Wolsey at Hampton Court in the 16th century, so if it makes you nervous you are in good company.

COLOURFUL BERRIES

The natural world uses bright colours for two reasons: to attract or to warn. In autumn hawthorn, blackthorn, spindle, rowan and many more put out bright berries as advertisements to birds and animals, who will eat them to survive the winter. Some, such as blackberries, are delicious for humans, but the display is put on for wildlife, not us. When the seeds inside are deposited on the ground in droppings they will stand a good chance of growing into new plants and thus the feeders help the host plants.

BINS NIGHT IS PARTY NIGHT

Rubbish collection can be extremely early so bins go out the night before. Word gets around and bins night is party night for urban wildlife.

In our street the first to arrive are magpies and crows who don't even need the encouragement of visible food scraps and will peck at closed bin bags, tearing them open. Foxes, rats, cats and gulls follow and before you know it the street looks like Glastonbury after the festival.

To avoid harming wildlife use strong bin bags (or better still put bags inside a bin) and ensure that any recycling is scrupulously clean. Tins can cause nasty cuts to any animal that shoves in its snout (or their whole head, these partygoers do not hold back). Empty and clean containers and pinch cans shut.

HOUSE SPARROW

The cheeky little house sparrow is quite comfortable around people, and despite numbers having declined markedly, we see them in town year-round. There is some uncertainty over the reasons for this decline, but one theory is the reduction in horse transport: sparrows were fond of the undigested grain found in horse manure.

House sparrows nest in holes in buildings, and are frequent users of our nest boxes, where they can rear up to four clutches of young in a year. The male will often take over once the chicks are hatched while the female prepares herself for the next brood. Both male and female are a soft grey-brown, but the males have grey cheeks and black throat and bib.

MAPLE

The maple family is large, including sycamores, the sugar maples of North America and our field maple, slightly misnamed as it makes a good urban tree. Today though we are looking for ornamental maples; smallish, brilliantly coloured trees planted as deliberate displays in what are often called Japanese gardens. They are often more like shrubs than trees and have delicate, pointed leaves with 5–7 lobes. Look at the individual leaves as well as the overall colours of the trees, many have intricate patterns in shades of red, orange and yellow. They are perfect for small gardens too, growing well in containers. Japanese maple (*Acer palmatum*) is the species to look for, 'Inaba-shidare', 'Burgundy Lace' and 'Orange Dream' are our favourite varieties.

SNAIL

As snails are nocturnal you must go out at night or in the early morning to watch them. They don't move fast so take the time to appreciate their beauty. Snails do not slither, but instead have a single muscular foot, which they use to propel themselves along in a series of pulses – really more of a hopping crawl. They need moisture to survive and in very dry conditions they simply go to sleep and that nap can, if needed, last up to three years.

Snails are born with those beautiful spiral shells, although at that point the shells are very soft and only harden up as the snail grows. They can heal themselves of small cracks or chips in the shell but need it to survive. They are different from slugs: a snail without a shell is not a slug, it is a dead snail.

DOGGIE DELIGHT

We can and should learn a lot from the natural world. Watch dogs at play. Have you ever seen anything that exemplifies pure joy quite this much? Whether he is running round the park with the wind in his tail, ears flapping or rolling round on his back in a patch of dust on the towpath, that is one happy dog. Apparently dogs have the same emotions as two-year-olds and they are expert at living in the moment. Interestingly, one emotion that passes dogs by is shame, which is fortunate given the range of doggie daywear on show in our local park. So go on, embrace your inner dog, and try just occasionally to really live in the moment and capture that joy.

ACORNS

Today people rarely forage for acorns but in the past they have been widely used for flour, the name coming from 'oak-corn' or 'ac-corn'. During the Second World War they were also used as a coffee substitute. Pigs love acorns but in urban areas they are more likely to be collected by jays or squirrels. Unwittingly, these are largely responsible for the natural spread of oak trees; forgotten caches of acorns can eventually grow into majestic trees. On a more fanciful note, the stalked acorn cups of the English oak make better pipes for fairies (assuming they have not heard about the dangers of smoking), while the stalkless cups of the sessile oak make better fairy cups and bowls.

CRACKED CAP BOLETUS

It must be International Mushroom Day as overnight a veritable troop of mushrooms (yes troop is the collective noun for mushrooms) have sprung up on the verge. These look like cracked cap boletus, so although they are not harmful, this is not the time to put mushroom soup on the menu. Their cap has a crazed appearance, with spongey, yellowish spores underneath. The slugs are obviously keen as many show signs of nibbling. What is amazing is how quickly they have come up. Triggered by warm, damp weather, the underground network of mycelia has just 'bloomed' into full-grown mushrooms. If you are interested read *Entangled Life* by the magnificently named Merlin Sheldrake. This fascinating book explores underground fungal networks. Honestly, you will never look at a mushroom the same way again.

JACKDAW

Hanging around the picnic tables in the park, we meet a bold jackdaw making a play for a sandwich. The silver sheen on his head and his pale eyes mark him out from his cousin, the crow. Jackdaws are clever birds that can mimic many sounds and even be taught tricks. They are unafraid of people, love making eye contact and, in the past, they were often kept as pets. The jackdaw is attracted to bright, shiny things (who isn't), which has given him a reputation as a thief. Like people, jackdaws often go grey as they age. Their cry is a harsh 'tchack', which may have contributed the first part of their name; 'daw' may derive from an Old English word denoting a smaller species.

12 OCTOBER

IVY

Look at almost any derelict building, dead tree, old fencing or piece
of waste land and you will see ivy. Then look at manicured gardens
or well-tended parks and you will also see ivy. Unconcerned about
sunlight or shade, wet or dry soil, ivy will spread wherever it is given
the chance, clinging to its host. Ivy is not necessarily responsible for
destroying its host; it is only destructive if the structure in question
is weak with pre-existing gaps for the aerial roots to penetrate, or the
weight of the ivy smothers its support. The leaves of mature ivy lose
their lobes and become ellipse-shaped. In autumn the mature plants
also bear the clusters of greeny yellow flowers and green, later black,
berries so vital to the survival of many bees, butterflies and birds.

13 OCTOBER

SPIDER'S WEBS

Even committed arachnophobes can see the beauty in a spider's
web. Our favourite moment is walking through the park on a foggy
morning with just enough sunlight to illuminate the dew on the
webs: crystal drops on spun silk. But you needn't leave home to see a
web, they are often spun across windows or in corners.

In reality a web is not a home but a trap; a sticky net spun to
catch insects for later consumption. Different spiders make different
kinds of webs: some are sheets, others tunnels or orbs. The sheet
web spider even makes a web around a hole in which she lurks, one
foot on a signal thread that will vibrate when something is caught on
the web.

14 OCTOBER

SPINDLE

The spindle grows more often as a shrub rather than a tree, with
several stems making an upright, bushy shape. The pale yellow
flowers are small, the leaves thin and narrow and, until this time of
the year, it is a fairly unremarkable, if perfectly pleasant plant. Then,
suddenly, in autumn, it has its moment of glory with pink fruits and
fabulously coloured leaves. The fruits are almost indecently bright,
and the four lobes split open to reveal an equally bright orange seed,
whose colour should clash horribly but doesn't. Don't be tempted to
eat them; they are poisonous to humans.

MOTORWAYS

Motorways may not seem an ideal environment for wildlife, but this is another one of those 'you might be surprised' moments. Our motorway network is huge and runs past or through some of our most precious habitats. The Highways Agency partners with Wildlife Trusts to look at ways of minimizing the impact. Most motorways are bordered with buffer zones of trees or scrubland, which can be havens for wildlife (at least those that don't mind traffic noise), and create useful corridors for pollinating insects, birds and small mammals. The M5 in Somerset is working with its local Wildlife Trust on a project to build dormice bridges (we so want to see these) to help reconnect isolated populations of rare hazel dormice.

As you drive down a motorway be on the alert for birds – although this should really be a passenger activity, the driver should keep their eyes on the road. This can be a great opportunity to spot raptors such as red kites and buzzards. Outside of town they have more room to soar.

ROACH

As you wander along the towpath keep your eyes peeled for shoals of roach. This large, silvery fish with reddish-brown fins is the most abundant piscine inhabitant of our canals. Their high tolerance for pollution and hardy nature may account for numbers. They look similar to carp but are more sociable and likely to be found in shoals often hanging out in shaded areas under trees or bridges. Unusually they will crossbreed with other species, most often chubb or rudd.

ORB-WEAVER SPIDER

A spider has formed a web across the window (lest you regard this as a slur on our housekeeping skills, may we remind you that most spiders take only 30–60 minutes to spin a web, so keeping up with them is a Sisyphean endeavour). Spiders love forming webs in windows as insects are attracted to the light shining through. At the moment the spider is sitting right in the centre, affording a glorious view of his delicate legs and the white cross markings on his brown body. The light may help the spider attract prey but it also useful for us in observing and identifying the little chap as an orb-weaver. These spiders usually hang out in the garden but this one seems to have come inside.

18 OCTOBER

LEAF COLOUR IN AUTUMN

If spring is pale green and summer dark green, then autumn is a positive riot of colour. In urban areas exotic imports combine with native species to make this one of the brightest times and, while spring blossom only lasts a few weeks, the brilliant autumn colours can remain on the trees far longer in a good year. The colours come from pigments, but the variations are influenced by the weather; cold nights make the leaves redder. The length of the spectacle also depends on the autumn weather; strong winds and downpours will send leaves to the ground, while in gentle weather the display can last well into winter.

MIGRANT BIRDS

'Nice place to visit but you wouldn't want to live there', appears to be the view of around half of Britain's bird population who regularly migrate each year. Some travel relatively short distances, but others cover thousands of miles in search of better food sources or warmer temperatures. That migration is definitely a two-way street. Swallows are well-known migrants overwintering in Africa and returning to us as a sign of spring. Less well known is that our resident blackbird and starling populations are boosted by migrants who come to Britain from Scandinavia and Eastern Europe to escape the punishing winters there.

Just how they navigate these long journeys is uncertain; some species use visual landmarks like rivers and coastlines, others the earth's magnetic field, a sense of smell or even just following older, more experienced birds.

20 OCTOBER

FLY AGARIC

The fly agaric is the enchantingly pretty fantasy mushroom with the white spotted red cap that looks as if it belongs in a fairy tale. However, looks can be deceptive and this is a mushroom to look at, paint and photograph, but not touch. Be careful, as behind that beauty it is poisonous and hallucinogenic. The fly agaric attracts and kills flies (hence the name), and in the past dried pieces of the cap were broken up and steeped in milk for use as an insecticide. The best time to see it is in autumn on heaths and commons, often close to birch trees.

21 OCTOBER

APPLE DAY

Apple Day was started at the Old Apple Market in Covent Garden,
London by community conservation charity Common Ground on
21 October 1990. Their website says: 'It was intended to be both
a celebration and a demonstration of the variety we are in danger
of losing, not simply in apples, but in the richness and diversity of
landscape, ecology and culture too.' Anywhere there is an apple tree
there can be apple day celebrations: apple, juice and cider tastings,
games, music, cookery demonstrations, apple identification,
anything that makes people aware of the apple trees in your area.

      Celebrations are usually held on the surrounding weekends.

commonground.org.uk/apple-day

EELS

Eels are underwater lurkers with big teeth, which can be slightly disconcerting but rise above it, they have a very interesting story. Eels are born in the Sargasso Sea, which sounds very romantic, but is in fact part of the North Atlantic off the east coast of North America. Those planning a life in Britain then undertake a journey that can take up to three years and arrive here as young elvers. They like dark, murky water close to banks (their eyesight is terrible, so the murk makes no difference to them) where they scavenge off dead and dying animals. They then spend up to 20 years living in freshwater rivers before returning to the Sargasso Sea to spawn and start the cycle again.

23 OCTOBER
SWEET CHESTNUT

The sweet or Spanish chestnut is not actually related to the horse chestnut even though their seeds look slightly similar. The sweet chestnut has very distinctive long, pointed leaves, which are sharply toothed and long male catkins that appear in spring. The autumn fruits are round and bright green, like those of the horse chestnut, but are far spinier. Inside each fruit there are two or three brown nuts that nestle neatly together. Either pick fruits which have opened or gently nudge them with a boot, the spines can penetrate almost any gloves. Peel away the brown shell and the bitter pith – you can eat the nuts raw, but their flavour is unpredictable. We prefer them roasted; just remember to make a small hole in the shell so they don't explode on the fire.

SQUIRRELS: RED, WHITE AND BLUE

Okay, we may be teasing about the blue ones but there are red, white and black squirrels. We once saw a red squirrel in a riverside park. Aside from the obvious, red squirrels are quite distinct from their grey cousins. In fact, colour is not a clincher, as some greys can look reddish-brown. Reds have tufted ears and the grey has a distinct white rim or fringe to his tail. Reds are also smaller (but then so are baby greys).

The whites are either albino or leucistic and are very, very rare – estimates suggest no more than 50 in the whole of Britain, but as this is a genetic condition they are as likely to be in town as the country. Black squirrels are the result of inbreeding between greys and American fox squirrels that have escaped from zoos and wildlife parks.

STARLINGS

Early autumn evenings offer the best chance of spotting a murmuration of starlings, as at this time of year they gather in flocks and if the flock is disturbed they will rise up in a glorious, cacophonous cloud of birdiness, swirling and making huge Spirograph patterns against the sky. To be classed as a murmuration there must be at least 500 birds, but murmurations can contain many thousands. It is an absolutely breathtaking sight. It is not certain exactly why so many birds roost together: perhaps for protection against predators, to exchange information about food sources or just to keep warm. Whatever the reason, observing a murmuration is a very special moment.

WALKING THROUGH LEAVES

Autumn is often called 'fall' as that is what the leaves of deciduous trees do at this time of year. These trees use the energy from sunlight and dropping their leaves helps them conserve energy during winter when light levels are lower. This is the time of year to deliberately shuffle your feet: there are leaves everywhere on the ground and walking through them is one of the joys of autumn. Pick a dry day but wear wellies as the lower layers will be damp and walking through leaves with dirty, wet feet is less fun. If the leaves are left undisturbed they will break down and eventually enrich the soil below; if you dig into the leaf litter you will find wetter, more compost-like layers further down.

ROOFTOPS

There is something particularly striking about the combination of stark city skylines and the freer forms of nature. Rooftop gardens are delightful – emerging from a lift on top of a tower block to find yourself in a green wonderland is always a surprise, but sometimes it is the small touches of nature that astound the most: birds roosting on a powerline or along a ledge outlined against the sky; strangely angled plants growing out of cracks in the concrete several floors up; a swarm of wasps leaving their colony under the eaves of the church; or the sound the wind makes as it swirls around the towers. Make a point of looking out from your office window now and then (you need that break from the screen) and see what you can see and hear.

DAYLIGHT SAVING TIME: FALL BACK

The moment the clocks go back marks the start of winter for many people; that extra hour in bed being the only obvious reward for darker evenings. But it needn't be all bad. The mornings gain a little extra light for a while and afternoon walks can become magical when they tip into twilight. Appreciate the sunsets and take advantage of the fact that you can star gaze without having to stay up half the night. Cities take on a different feel in winter, nature is still there, just often less obvious.

29 OCTOBER

CANADA GOOSE

Canada geese were first introduced into St James's Park in the 17th century and have never looked back. They are comfortable with urban living, and you find them in most parks at best loudly begging for food, often demanding with menaces. Think of them as lakeside avian muggers. They have brownish bodies and black heads and necks, with a white patch at the throat. Long-lived and prolific breeders their aggressive territorial attitude can cause problems for other waterfowl. Technically speaking, they are a migratory species flying north for the winter, but in practice many are so happy in the UK they stay year round. When they do take flight, their iconic 'V' formation flight pattern is well worth watching.

BAT

Bats have always suffered from rather a sinister reputation due to
their starring role in too many horror films. Seen close up, bats
are more like little furry mice with rather sweet faces, somewhat
perplexingly dressed in punkish leathers. There are some 18 species
of bat in Britain, with the pipistrelle the most common. They are
mammals – the only mammal capable of controlled flight – and
are true acrobats. Bats hibernate during the
coldest months, so you are unlikely to
spot them in winter.

31 OCTOBER
CHURCHYARDS AND CEMETERIES

According to Marvell (that's the poet not the comic) 'the grave's a
fine and private place', and this description can be extended to the
surrounding churchyard or cemetery, both of which are favourite
spots for an urban nature ramble. Ancient or modern, they are often
planted with magnificent trees (traditionally but not exclusively
yews), and the lichens on gravestones are beautiful and may be as
old as the stone itself. Churchyards are green oases in the middle
of a city, some of which have sat undisturbed in the same space for
hundreds of years. Look out for mosses and lichens, wild flowers and
grasses growing around the headstones enjoying this contemplative
space as much as you do.

NOVEMBER

Much of nature adapts and copes with city life; some species positively prefer it. Plane trees still stand proud, holding onto their leaves and fruits well into winter, squirrels and foxes are adept urban foragers and birds take advantage of the rich pickings on thoughtfully stocked bird tables. Yes, it's cold and dark but fog and mist and rain are all pretty incredible viewed from a position of comparative warmth and dryness (think thermal socks and a good brolly), and the long nights make star gazing so much easier.

Green is the plane-tree in the square,
 The other trees are brown;
They droop and pine for country air;
 The plane-tree loves the town.

From 'A London Plane-Tree' by Amy Levy

1 NOVEMBER
DOG ROSE HIPS

The dog rose's pale pink summer flowers turn into pinky red hips
in autumn. These usually hang in clusters and can be used to make
syrup or tea. The syrup is a particularly good source of vitamin C,
easily outstripping oranges. It was widely used during the Second
World War and after the war a factory in Newcastle paid local
children 3d (about 35p today) per pound of hips. This may sound
like exploitation, but the children knew that they could harvest a
second crop from these wonder-fruits; each hip is covered with
protective hairs, more effective at causing irritation than any shop-
bought itching powder.

2 NOVEMBER
HORNBEAM

These city trees are frequently overlooked. Their leaves are similar
to beech, but they have irregularly toothed edges and very prominent
veins, which are clearly visible even at some distance. Traditionally
they were grown in and near cities to provide fuel; the wood is very
hard and the resultant charcoal burns efficiently and economically.
You may find hornbeam hedges where the dense foliage creates a
good barrier. At this time of year, the trees will be hung with three-
lobed bracts, looking like little leaves, each encasing a small nut.

BEETLE BUMPS

Those piles of rotting branches, wood chips and leaves in parks or
in school grounds are not examples of a gardener's neglect, they are
a valuable contribution to habitat management and provide a great
spot for a bit of insect observation. Think of them as bird tables for
beetles, providing ample food and a space to overwinter. Insects will
lay eggs in the logs and in time the grubs will munch down on the
rotting wood. Look closely and see who has made a home there but
don't disturb the pile.

THE BEAUTY OF FALLEN LEAVES

Autumn leaves look amazing on the trees, but they can be just as
wonderful when they fall to the ground. One of the great pleasures
of autumn is admiring fallen foliage. Looked at close-up, some of
the colours on individual leaves are truly astounding: buttery yellows,
glowing oranges and the deepest imaginable crimsons. Leaves are
easy to press (you can use acid-free or waxed paper, but we find
ordinary newspaper works perfectly well) and can be used to make
decorations or pictures. Finally, find a lake or patch of still water
overhung with trees and take time to appreciate the beauty of the
floating leaves, both on the water and as part of the reflection.

5 NOVEMBER
STARS

Today is not the best time to go star gazing: the sky will certainly
be filled with wild colour and starbursts, but this is all due to the
fireworks traditionally let off at this time of year. It is very pretty and
a lot of fun but for a more serene celestial beauty give the night sky
the same amount of attention at other times of the year. Pick your
moment. Winter is better than summer to avoid any haze, and a
moonless night will let you focus more on the stars themselves. Get
as high as you can: a loft window or rooftop is good. Keep all lights
off (even that phone, put it away) and let your eyes acclimatize to the
dark. Now just look at the stars, there are billions of them out there;
if you are very lucky you will see a shooting star, don't forget to wish.

6 NOVEMBER
ASH TREES AND THEIR KEYS

The mountain ash is a rowan, while the common or European ash
is a very different tree, one of our large, noble natives with bark that
darkens and develops deep fissures as it ages. The leaves are made
up of 9–13, long, pointed leaflets on a single stem. These appear in
late spring; the small male and female purple-green flowers growing
first. At this time of year, the trees are covered with untidy clusters of
single-winged fruits or keys, which remain on the tree long after the
leaves have fallen.

Ash dieback is a fungus that originated in Asia and was
first recorded in the UK in 2012. It was unwittingly introduced on
infected saplings, but the spores travel in the wind and the disease
has spread naturally and rapidly. The European ash has no natural
defence and at the moment it is estimated the UK could lose 80 per
cent of its ash trees.

7 NOVEMBER
FOG AND MIST

Smogs and pea-souper fogs were common in cities when the combination of heavy industry and coal fires filled the air with pollution. They often reduced visibility to a few metres but the Clean Air Acts of 1956 and later have largely made these a hazard of the past (some things do get better). Mist and fog are formed from water droplets near the ground, which clear away as the sun rises and warms the air, turning the water into invisible vapour. Radiation fog forms when heat rises from the ground into colder air above; while advection fog is produced by warm, moist air passing over a cooler surface, often water. The difference between fog and mist is simply one of intensity, the point of separation being visibility of more or less than 1,000 metres (3,280 feet).

8 NOVEMBER
FLIGHT

Like birdsong, flight patterns are a guide to identifying different bird species. There are different styles of flight: hovering, gliding or flapping. Some small birds have a bounding flight, where short bursts of flapping are interspersed with periods when the wings are folded, and the bird bounds along. Watch how the birds move silhouetted against the sky. Are they alone or in an almost military 'V' formation or more like a flock? Some birds spend the majority of their time in flight: eating, mating and even sleeping on the wing; they can spend weeks in the air catching ten-second powernaps. The principles of how flight works is similar in birds and aeroplanes: lift, drag and thrust.

TURKEY TAIL

The turkey tail is an impressive fungus (now that is a sentence we never thought we would write). It is not huge, but the concentric rings of colour (browns, greens, purples and blacks with a pale outer ring) make it easy to identify as it does look very like its namesake. Turkey tail used to be collected and used as table decorations (our ancestors had a more laissez-faire approach to health and safety) and even, so we have read, to decorate hats. It often grows in layers up the sides of decaying trees, or on dead logs where it helps to break down the dead wood, turning it back to soil. It is a tough fungus and, although we have heard it can be chewed like gum, we have never tried it and almost certainly never will.

10 NOVEMBER
PUDDLES

As a youngster, splashing through puddles was a favourite activity, but before you reconnect with your inner child take a moment to examine that miniature lake. Puddles are transient – they pop up after rain and can be deep as the sea (well maybe not quite) or a shallow mirror of water across a pavement. Look at the reflections: that framed shot of the sky or overhanging trees may look quite different to the image held in the water. Here is another great photo moment: small birds and insects love puddles and use them for bathing or drinking and plants will grow up around the long-lasting puddles. And that puddle may house a mini world of tiny organisms.

11 NOVEMBER
SAINT MARTIN'S SUMMER

It is unseasonably warm, although with climate change is anything unseasonal any more? Seasonal temperatures are a guideline not a rule. This type of weather used to be referred to as an 'Indian summer', a name adopted from the Americans in the 19th century. It apparently referred to unusually warm weather occurring in late autumn, which allowed Native Americans to extend their hunting season. Nowadays we prefer the traditional British name of Saint Martin's summer. Saint Martin was a soldier best known for cutting his coat in half to give half to a beggar suffering from cold. He is the patron saint of soldiers and beggars, and his saints' day falls on 11 November or Armistice Day, right at the end of the harvest season.

When we get one, a Saint Martin's summer is a perfect time for an autumnal drink in a pub garden. You can bask in the sunshine while admiring the autumn colours, very much the best of both worlds.

BEECH NUTS

The fruit of a beech tree is called mast, which comes from the old English word *mæst*, traditionally meaning the fruit of beech, oak, chestnut or other forest tree used as food for pigs. Beech nuts start out soft and green, later turning brown with a tough, hairy casing. This splits into four parts revealing two shiny, brown triangular nuts. Every 3–4 years there is a 'mast year' when the trees produce particularly large quantities that litter the ground. Some town councils avoid planting beech trees as they regard the mast as a hazard. These are beautiful trees; surely a solution would simply be to allow pigs to roam our streets? Answers on a postcard please ...

RIGHT TO ROAM

Oh the moors are rare and bonnie
An' the heathers sweet and fine
An' the roads across the hilltops
Are the people's — yours and mine.

This rousing anthem was sung by the 10,000 people who processed up Winter Hill in the West Pennines in 1896 in one of the largest acts of mass trespass in history. Their actions recognized the huge role nature played in restoring physical and mental health to city workers labouring in cramped and polluted conditions. Sadly, their right to enjoy the open air clashed with landowners' right to privacy, and from 1892 to the 1930s a series of pitched battles ensued. Landowners would close off footpaths that had been enjoyed by generations and left-wing activists would organize mass protests.

Today the Countryside and Rights of Way Act 2000 gives access rights to limited parts of Britain and surveying authorities are required to keep up-to-date and definitive records of public rights of way. By contrast, in Scotland, there is no law of trespass although you must not cause damage to property.

The organization Right to Roam (righttoroam.org.uk) petitions for legal change and also promotes the use of peaceful and responsible acts of trespass to highlight the issue.

ROWAN TREES IN AUTUMN

In spring the rowan or mountain ash sports feathery, pale green leaves; in autumn these darken, providing a perfect backdrop for bunches of bright red or yellow berries. The leaves then turn every shade of orange, red and gold before dropping but the berries remain, providing the birds with much-needed winter feasts. Individually they are not particularly striking trees, but lining residential streets they provide avenues of welcome colour. The berries are poisonous to humans if eaten raw, but can be made into a sharp and rather delicious jelly.

15 NOVEMBER

ROOK

Rooks are very sociable, if you think you see one by itself look again, it is probably a crow. In the local park we often see them hanging out with jackdaws round the café's outside tables, ever hopeful for a dropped sandwich or unguarded chip. Their pointed white beaks look as if they have been dipped in a tin of emulsion, and if the sun is shining their black feathers often have a blueish-purple sheen. They build their nests close together in colonies known as rookeries, a term since extended to refer to city slums.

BLACKTHORN BERRIES OR SLOES

Blackthorn produces distinctive blue-black berries, which are a surprising bright green inside. These are not berries to nibble as you pick – they are best made into sloe gin (or vodka, whichever is your preferred tipple). Unlike blackberries, sloes should be picked after the first frost as they will then release their juice more readily. If you are worried someone else will race you to the harvest, you can always pick the berries as soon as they are soft and put them on trays overnight in the freezer. The rounded trees grow up to 4 metres (13 feet) but are usually shorter, with oval, green leaves and very dark bark. Beware of the sharp thorns – we wear those leather gauntlets – yes they do make you look a bit ridiculous, but that is better that than bleeding all over your harvest.

17 NOVEMBER

WOODLOUSE

Lifting up a stone there is a scurry of escaping woodlice: those tiny little armadillos of the insect world. In fact, they are not insects at all, but a type of crustacean that has evolved to live on the land. It is possibly due to their marine origins that they love dark, damp spots and are happiest at night when light levels are low. There are some 250 dialect names for woodlice in Britain, including woodpig, chiggywig, cheesy bug and slater. Their carapace is divided into seven parts like tiny plate armour, and is shed every two months, losing first the rear half, and then once that has regrown, the front. They can be a touch smelly (stinky bug is another title) as they excrete an ammonia-smelling substance through their shells.

FLOWER ARRANGING FOR AUTUMN

If you are lucky enough to have a garden, one of the treats of summer is to pick a bunch of flowers for the table. This needn't stop in autumn – you just need to adapt your arrangements. Many flowers look just as good or even better as they fade: hydrangeas, alliums and the fluffy seed heads of some clematis. Quaking grass, oat grass and pennisetum all improve their looks as autumn approaches, and the great thing about grasses is that you only need to pick a few stems to make a great impact. Fill the bulk of the vase with pretty stems and use the flowers and grasses to make the statements.

HAWTHORN BERRIES

The name hawthorn comes from the Old English *haw* meaning 'berry' or 'hedge', the thorn part will be obvious to anyone who has come into close contact with the plant. The haws or berries start out green, turning bright red as autumn progresses. Traditionally, it was thought they should not be picked until after the first frost, but in practice they are ready to pick once they have turned red. Never strip a tree of all its haws as they will remain on the branches until late winter and are a favourite food of many birds. The berries can be eaten raw but are best made into jelly; you should not eat the seeds.

HERBALS

Old herbals – books containing descriptions and uses of plants –
are entertaining to read and you may find some interesting uses
for herbs or plants you buy or are growing. The two best-known
are John Gerard's of 1597 and Nicholas Culpeper's of 1652–3.
John Gerard had a large herb garden in London but much of his
book was plagiarized from earlier works. However, his writing was
entertaining, and the book was very popular. Nicholas Culpeper was
a physician and aimed to improve the health of the nation, treating
the poor for free and promoting basic remedies. Both have useful, if
surprising recipes for conditions such as baldness, giddiness, obesity
and hangovers.

LAKES AND PONDS

Most decent-sized parks will include a pond, and sometimes even a
boating lake. Village greens have them, as do heaths and commons.
Who didn't grow up feeding the ducks? They are great meeting
places, perfect for picnics and beloved by local wildlife. Think
beyond waterfowl though, there is a huge diversity of fish, insects and
amphibians in and around the water, and the plants you can see there
will be quite different from those on drier ground. Water lilies are a
particular favourite of ours, although sadly not at this time of year.

OLD MAN'S BEARD

Unattended trees or hedges are often draped with a wild clematis that has two common names: traveller's joy or old man's beard. The small flowers are a greenish yellowy white and smell pleasantly of vanilla, making walking along paths overhung with the plant a joy. It is in autumn though that the plant becomes most noticeable, with fluffy seed heads resembling a large but scruffy beard. As children we were told that abundant old man's beard foretold a harsh winter, but we can't remember whether this was true; by the time it snowed we were too excited to remember what plants had been doing in autumn.

BROWN OR GREEN

In every city there are areas where human activity has paused and nature has taken over: building sites on hold, plots waiting to be developed, areas round factories. Anywhere left unattended by people gives nature an opportunity to move in. This land is often referred to as 'brownfield', as if it is worthless and devoid of living things, but nothing could be further from the truth. Free from human intervention, wild flowers, invertebrates and insects flourish. Naturalists have written about it: Richard Mabey called it *The Unofficial Countryside*, Stephen Moss, *The Accidental Countryside*. Many of these areas are fiercely fenced and, while we would obviously not encourage or condone trespass, if you can gain access to them, prepare to be amazed at the diversity you discover.

DON'T FIGHT NATURE IN THE GARDEN

People fortunate enough to have a garden learn, very early on, that fighting nature is a battle they will lose. Most garden plants started off life in the wild, and they will only thrive in our gardens if we can provide them with suitable conditions. This means working with nature to create the perfect balance for your plants. It may seem like a quick solution to zap unwanted visitors with a spray gun, but your garden will be healthier and more interesting if you encourage natural predators. Birds will eat unwanted slugs and snails, and most bees, butterflies and other insects will help your plants rather than harm them. Think before you kill.

25 NOVEMBER
REWILDING

There is more to rewilding than simply ignoring the land. The aim is obviously to let nature take over, but sometimes the thugs need controlling and the weaker species need a helping hand. Two of our favourite books are *Wilding* by Isabella Tree and *The Garden Awakening* by Mary Reynolds. *Wilding* is the story of allowing nature to return to a once-intensively managed farm in West Sussex; while Mary Reynolds actively encourages the wild into her garden. Both are on a considerably larger scale than most urban gardens, but they make inspiring reading.

26 NOVEMBER
SUNRISE

One of the advantages of the short days of winter is that it is comparatively easy to get up in time to watch the sun rise. Depending on the conditions, a glow of light may appear 20–30 minutes before the actual sunrise. You won't be able to see the sun because it will still be below the horizon, but its light will be visible along the curve of the earth. In photographic terms, the blue hour occurs just before sunrise (and after sunset) when the light often appears cool, while the golden hour is just after sunrise (or just before sunset) when the low position of the sun creates a warm glow. In practice these periods may only last a few minutes rather than a full hour, but they can be very beautiful.

NATIONAL TREE WEEK

The dates vary slightly from year to year, but National Tree Week always takes place about now, and is a perfect incentive to get out and look at trees. Our autumn colour may not be as brilliant as that of the north-east United States, but look out for individual trees in woods, parks or gardens. It also marks the start of the tree planting season, and all over the country conservationists and volunteer groups plant thousands of trees, hedges and orchards in both cities and the countryside. There are brilliant activities for adults and children alike. For details see the Tree Council's website (treecouncil.org.uk).

28 NOVEMBER
RAIN

You do not always need to leave the house to connect with nature. On wet days we like to take some time out, make a cup of tea and sit by the window watching the rain. The pattering sound is soothing, and even if it is lashing down you can relax, you are inside. Watch the water as it runs down the glass. In the 18th century, gentlemen at the London club Whites would bet on which raindrop would reach the bottom of the pane first. You may not want to put money on it, but the patterns are beautiful: drops, streaks, whirls. If it is dark the rain around the streetlights will shimmer and glow. It is a kinetic art installation available to us for free.

ROUNDABOUTS

Don't get distracted while whizzing round a roundabout on your
bicycle but if you are waiting in traffic or just walking past we would
highly recommend a closer examination. Either by accident or
design, many roundabouts have become havens for local wildlife.
One near to us would make Wordsworth proud with its perfect
golden host of daffodils in spring. Some roundabouts are community
gardens, or carefully maintained as wildlife meadows by the local
councils, attracting a huge diversity of flowers and insects. Others
have been ignored for decades but have taken it upon themselves to
become secret nature sanctuaries. Isolated by a sea of traffic their
residents are undisturbed by pedestrians and predators alike. There
was a roundabout in Norfolk that sheltered a flock of feral chickens
for years.

THISTLES

There are a number of different thistles; the spear thistle is often
regarded as the most likely candidate to be the emblem of Scotland,
but no one is certain. As writer and naturalist Geoffrey Grigson
points out, the Kings of Scotland were rulers and warriors rather
than botanists, and they chose the thistle for its splendid purple
flowers and no-nonsense prickles. We have included thistles today in
honour of St Andrew's Day, the patron saint of Scotland but, by now,
most will be past their prime with the purple flowers transformed
into fluffy seed heads. Thistles can be found on any untended patch
of ground; wildlife loves them, with insects living among the safety
of the spines, butterflies enjoying the nectar in the flowers and birds
feasting on the seeds.

DECEMBER

We would like to wake up to snow on Christmas morning and see our cities transformed into fairy-tale wonderlands but, in reality, Robert Bridges' words are probably the nearest we are likely to get to that dream. However, there is still much to admire in the natural world and, as an antidote to the human-inspired rush of the month, it is worth taking time to enjoy spectacular sunsets, feisty robins, extraordinary fungi and the brave evergreen trees that defiantly keep their leaves. There are illuminated evening walks to enjoy and, whether you celebrate the Winter Solstice or Christmas (or both), now is the time to bring nature into the house with garlands of holly, welcoming wreaths and, of course, mistletoe.

———

When men were all asleep the snow came flying,
In large white flakes falling on the city brown,
Stealthily and perpetually settling and loosely lying,
    Hushing the latest traffic of the drowsy town;

From 'London Snow' by Robert Bridges

1 DECEMBER
HIBERNATION

We have often said we would love to hibernate over winter, but humans just have not evolved that way. Hibernation is when animals build up reserves of body fat and tuck themselves away in a state of biological stasis: their heart rate slows and they can survive the winter without eating. When bats hibernate, their heart rate drops from 1,000 beats a minute to just 25, and they breathe just once every two hours.

In Britain only hedgehogs, dormice and bats hibernate. Other creatures have different ways of coping with the cold temperatures and food shortages of winter. Squirrels, foxes and badgers just slow down: sleeping up to 20 hours a day and venturing out to feed in the few hours they are awake. Other creatures spend winter in a different state – as an egg, caterpillar or pupa.

2 DECEMBER
CHRISTMAS DECORATIONS

Everywhere you look at this time of year you will see pictures of houses (usually large) decorated with swathes of greenery on every available surface. This is fine if your house has lots of trees in your garden, or you live next door to a forest (or have a florist on speed dial). Urban dwellers will need to cheat a little. First, use either very small or very large vases. Use the small ones as statements in strategic points, filled with tiny flowers or beautiful little stems. Fill the large vases with branches (dead or alive) and decorate with lights and baubles. Finally, collect long strands of ivy and drape them over select surfaces, with candlesticks, decorations and more lights. Check with the garden owners first but it is unlikely anyone will mind you collecting a few strands of ivy as long as you are careful.

3 DECEMBER
CANDLESNUFF FUNGUS

In the park today we spotted candlesnuff fungus on some dead and rotting logs (it is particularly keen on beech, but can be found on other woods). Looking like dead men's fingers or miniature antlers (another common name is stag's horn fungus) the whitish grey protuberances rise up eerily from the logs. It is small – only up to 6 centimetres (2½ inches) high but it is quite common so keep an eye out in winter. If you give the fungus a very gentle flick it will puff out spores like a gentle wisp of smoke from an extinguished candle. It isn't edible (it's not poisonous, just tough and unappetizing) but does contain anti-viral and anti-carcinogenic properties, so there is potential one day for this little fungus to do great things.

4 DECEMBER
CHILDREN'S FANTASY

There are many fantastical worlds created in adult stories but if you really want to be swept away we think there are certain children's books you should read. Most obvious is the Harry Potter series. Leave aside the wonders of moving staircases and flying broomsticks and hunt out the nature instead: a willow tree with real personality, plants that scream and shout. And that's before you discover the wonderful menagerie of extraordinary birds and animals. Narnia has a magic tree, winged horses, unicorns, talking animals and one of the best descriptions of snow. *The Sword in the Stone* is the first part of T. H. White's Arthurian series, but in it Merlin gives most unusual nature lessons – you will discover exactly what it feels like to be a fish and you'll also meet a very special owl who talks *and* sulks.

CHRISTMAS WREATH

The origins of celebratory wreaths go way back. The Celts made them to mark the solstice and Ancient Greeks and Romans to recognize status or achievement. Using wreaths to mark Christmas began in the 1500s when wreaths incorporated candles to symbolize the light that Christ brought into the world. Today they are usually hung on your front door during Advent and a wreath is an invitation to bring good fortune and the spirit of Christmas inside. The circle shape symbolizes eternity, the greenery used tells of growth and everlasting life, and seeds, fruits or pine cones are a nod to rebirth and the coming of spring. You can pick up a pretty wreath at your local garden centre or supermarket but where is the fun in that? Try making your own.

6 DECEMBER
PEBBLES

There is something supremely tactile about pebbles. Worn smooth by the action of the river, they cry out to be picked up and handled. Gather them as you wander along the river at low tide or try just sitting down in one place and consciously examining those within reach; look at them properly, see how they change when wet or dry. What do you think they started out as? There are pebbles from bits of rock, streaked or veined with quartz, tiny flints, glass pebbles that may have started life holding some boatmen's tipple, or pieces of brick or concrete from long-demolished buildings. Use them as paperweights, make a collection in a glass bowl or scatter sharp ones around the base of your pot plants to deter slugs. Technically, a pebble is larger than a pea and smaller than a tennis ball, outside that it is a stone or a rock or even a cliff face.

7 DECEMBER
FLYOVERS

Walking under an urban flyover is a lesson in the resilience of nature. This is a very hostile environment: noisy, with poor air quality and little in the way of natural light and water. In many cases the natural world has left the area well alone, although you may occasionally see pigeons, spot some stunted buddleia or hear the scurry of rats. In recent years, people have tried to bring nature in artificially with beautiful murals such as the woodland wonderland painted under the Marylebone flyover, or the stunning ribbon of 50 colourful aluminium flowers waving under the M8 near Glasgow. There are also projects to improve the environment with green screens of ivy and other hardy plants.

TALK TO THE TREES

Trees can live for hundreds of years and witness so much. An oak reputedly sheltered the future Charles II fleeing the Battle of Worcester in 1651, and a sycamore witnessed the death of 1960s popstar Marc Bolan in 1977. Imagine if you could talk to them. What would they say? Many people, our King Charles III among them, do talk to trees claiming that it helps them think things through and encourages the tree to grow. Literature abounds with trees that talk back: the fabulous Treebeard of Fangorn Forest in *The Lord of the Rings* or the Fighting Trees in *The Wizard of Oz*. The wizard Merlin is turned into a tree in Arthurian legend. There is a current theory regarding the wood-wide web, which says that trees communicate and share water and nutrients via huge underground fungal networks.

DRIED FLOWER ARRANGING

Dried flower arrangements are a boon for the less-than-diligent home-maker (don't judge), as there is no need to worry about topping up the water, or having them die on you. This time of year, when everything starts to become a bit hectic, they can be a lovely and low-stress touch of nature in your room. Go for a mix of colours and textures; include some feathery grasses, seed heads, teasels or sculptural rushes. You can air dry your own flowers, which will take three weeks or so, but most florists have dried flowers readily available.

10 DECEMBER

STREETLAMPS

At night, for varying lengths of time depending on the season, rural areas are dark. Properly dark. This rarely happens in urban areas. Plants, birds and animals are all affected by the circles of artificial light that streetlamps create. The light attracts clusters of insects, which makes feeding easier for some bats, but is to the detriment and imbalance of the insect populations. Night-flying pollinators, who are the most effective, are diverted from flowers, who suffer in turn and then pass the imbalance on to day pollinators. Moths, who navigate by the moon, are disorientated and light-averse bats have a reduced territory.

11 DECEMBER
SUNSET

The most spectacular sunsets (and sunrises) are often seen during
the cold winter weather. There are various reasons, which combined,
give the perfect conditions at this time of year. Light is made up of
different colours: blue and green light waves are short, scatter more
easily and are filtered out and lost when the sun is low; red and
orange light waves are longer and can travel much better. In winter
the sun is low on the horizon for longer, allowing these colours to
build and intensify. On days when humidity is low, the colours also
appear much clearer, unlike the hazy sunsets of a summer's day.
Finally, a few clouds may break up the bands of colour and create an
even better display.

12 DECEMBER
COCKROACH

It is an urban myth that the cockroaches will still be standing
when the rest of the planet self-destructs. They are tough but not
that tough, and although more resilient than people, they are
not indestructible. That said, a cockroach can live a week without
its head. Most people, us included, are a little squeamish about
cockroaches – they are pretty ugly (something about those spiny legs
and the scuttling), spread disease and you really don't want them
in your house. However, they are an important part of the chain of
life, cleaning up dead and decaying matter and breaking it down
as nitrogen to enrich the soil. You have to applaud them for sheer
endurance.

TEAL

These are Britain's smallest ducks, only 35–38 centimetres (13–15 inches) long. The males have a chestnut-orange head with a striking green eye patch outlined in yellow. Their bodies are dark grey with a white stripe above the wing. Females are small, greyish brown and easiest to identify when seen with a male. Both have a bright green patch or speculum and white bars on their wings. The collective noun for teal is a 'spring', as they are able to take off vertically, as if they have springs on their feet.

14 DECEMBER
CHRISTMAS LIGHTS

Now is the time to wrap up and go and visit the Christmas lights. Not the ones in the city centre, although these are lovely too, but an illuminated trail in a local park or botanical garden that mixes nature with art. Lots of places do this now, so have a look in your local newspaper or community website to find one near you. This is a magical spectacle. Trees are illuminated in different colours, lights cascade through bushes or scamper across water and music plays. You can see yew domes festooned with fairy lights, follow a trail though a fire forest or see deciduous trees with their leaves replaced with lights. The nights close in early this time of year so as darkness falls, go out, grab a mulled wine and enjoy.

MOUSE

'There's a moose loose about this hoose.' Your first intimation that you may indeed have a mouse loose is likely to be droppings, small black pellets about the size of a grain of rice. A lot of droppings – a single mouse can produce 50–70 a day.

There are mice in both town and country, but a German study found that city mice are more streetwise and better at problem-solving, having adapted to a more varied environment. The house mouse, the more common urban species, has grey-brown fur, big ears and eyes and a pointed snout, and their feet are much smaller than their country cousins.

It is myth that they love cheese (although they will eat it) – their real passion is peanut butter.

16 DECEMBER
TOWPATHS

Originally designed for the horses, or sometimes people, that pulled the barges transporting freight along British canals, the towpaths that run through cities are a great place for some urban nature spotting. These green corridors are home to a wide variety of wildlife. There is the tantalizing possibility of seeing water voles and even otters, herons and kingfishers, and you will certainly meet ducks, geese and other waterfowl. The lush variety of land and aquatic plants attract damselflies, bees and dragonflies. Walk the same towpath at different times of the day and seasons of the year to vary the experience. See the water clouded with morning mist or sparkling in sunshine and note down what you spot in your diary.

17 DECEMBER
HOAR FROST

Hoar frost is the elegantly-hairy icy frosting on branches and grasses in the local park, or indeed, on traffic lights, drain grates or car windscreens and wing mirrors. Everything is transformed into a thing of beauty by a coating of hoar frost. It comes about when the water vapour in the air comes in contact with surfaces that are below freezing, and freezes instantly; this happens repeatedly and layers are built up, forming the icy crystal feathers we know as hoar frost. Now is the time to get out your camera (or your sketchbook) and try to capture the moment.

18 DECEMBER

ROBIN

Everyone loves a robin. It has been voted Britain's favourite bird
and Christmas isn't Christmas without a robin, whether it is in your
garden or on a Christmas card. Our robin is a frequent garden
companion, hopping about the flower bed as we dig, ever on the
lookout for a worm. They are quite territorial too, so you will often
see the same robin over a longish period. Robins are unusual in that
they will sing at night; we have seen a little robin by the streetlamp
singing its heart out. In fact, that nightingale that apparently sang in
Berkley Square was probably a robin, as you are very unlikely to hear
a nightingale in the city.

19 DECEMBER

ECLIPSE

When the moon moves between the earth and the sun, temporarily
blocking the light, this is a solar eclipse. The moon moves slowly so
this happens gradually. When the sun is completely blocked, this is
called a total eclipse. There are between two and five partial eclipses
a year but a total eclipse is rare and often visible from only one spot
on earth, making them an event. And you need to pay attention – a
total eclipse will last a maximum of seven and a half minutes. The
next total eclipse visible from Britain is predicted for 2090. A lunar
eclipse is when the earth is between the sun and the moon and the
moon lies in earth's shadow. The moon does not disappear but turns
a deep red.

20 DECEMBER
HOLLY

Holly is instantly recognizable with glossy, prickly leaves and, if
you are looking at a female tree, bright red berries. An abundance
of berries is meant to predict a harsh winter but we never trust
this; the fruits of a tree are dependent on the weather during the
previous seasons rather than those to come. The Celts regarded it
as protection against evil spirits and thunderstorms; the latter may
be true as the dense foliage and particular leaves make them good
lightning conductors. Holly has always been part of the winter feast
of Saturnalia, which the Christian Church absorbed into Christmas
rather than trying to prevent – the permanent greenness of holly
symbolizing immortality and the berries and prickles representing
the suffering of Christ.

21 DECEMBER
WINTER SOLSTICE

Depending on whether you are a glass half full or half empty sort,
today is the longest night or shortest day of the year. There are
some eight hours of daylight but a full 16 hours of night-time in
which to carouse or sleep. Our ancestors always marked the solstice,
considering it a magical time and a precursor of rebirth. The word
itself comes from the Latin *solstitium* meaning 'the sun stands still',
because the sun seems to stop in its path before changing direction.
At this point the earth is tilted on its axis at the furthest point from
the sun, and the sun is at its lowest point in the sky. In astronomical
terms the winter solstice heralds the start of winter although
meteorologically speaking it is about halfway.

22 DECEMBER

SNOW

The first fall of snow divides the adult world: 'The buses won't run', 'It will turn to slush by lunchtime,' or 'Oh, it's so pretty!' We, of course, fall into the last category, regarding snow as a treat rather than an inconvenience. In many urban areas it will quickly turn to slush, so the trick is to leap out of bed, get dressed as quickly as possible and go out and enjoy it. Admire the beauty, build snowmen and have fights if you wish, but then look at the individual snowflakes. Each one has six points and its pattern is unique. They form when water freezes round a minute particle of dust or pollen to form an ice crystal. As this falls to earth, it collects more ice crystals, which create a symmetrical and intricate pattern.

23 DECEMBER

RAILWAYS

Always look out of the windows of trains. From lush trackside forests to a single, spindly buddleia forcing its way out from under the sleepers, nature abounds around railway tracks. We have seen fox cubs playing in the sunshine at the Shacklegate Interchange and pipistrelle bats roosting in tunnels. Under railway bridges bindweed and ivy twine around thick concrete pillars colourfully daubed with graffiti in a perfect nature-meets-art moment. There are always thousands of blackberries alongside the tracks, but curb your enthusiasm here as rail companies routinely spray pesticides and this bounty is best avoided. These green corridors through cities form havens for all sorts of wildlife who seem undisturbed by the passing trains.

24 DECEMBER
MISTLETOE

If you see what looks like a very messy green nest in a tree, look again – it is probably mistletoe. The clumps can be up to 1 metre (3¼ feet) across, with pale green leaves growing in pairs like little wings. There are tiny greenish-white flowers early in the year followed by the all-important berries around Christmas. And why do we kiss beneath it? The Celtic Druids regarded it as a symbol of life because the berries grew at the coldest time of the year; Odin's son was resurrected after being shot by a mistletoe arrow; and around the world it is regarded as a fertility symbol. Depending on which tradition you follow (or who you are with), a single kiss is allowed or one for every berry.

25 DECEMBER
GIVING BACK TO NATURE

Christmas is not just for people, although we do enjoy it. It is a time for giving and now is the moment to give back to nature in thanks for all the pleasure you have received over the year. Put out some extra titbits in the bird feeders, go out and make that small hole in your fence to enable the hedgehog highway, plan to rewild a part of your garden or sign up for a river clean-up. Even just picking up some rubbish on a path will help make a safer environment for wildlife.

Think of yourself too – sometimes a cheeky Christmas present from yourself is just what you need, and bound to be welcome. Some new wellies? A pair of binoculars? A new sketchbook? What would you like to help you enjoy nature just that bit more next year?

BOXING DAY WALKS

Boxing Day is traditionally the time when the rich 'boxed up' gifts for their servants and a holiday when said servants got to go home and visit family to share or show off those gifts. Today Boxing Day is a time when, surfeited by overeating on Christmas Day and perhaps a little over the intense family time, we often choose to go out for a good, bracing walk. In town your local park or common beckons or perhaps just a few streets near to your home. Take the time to really look at your surroundings and enjoy what nature has to offer. Marvel at the Christmas lights threaded through trees in people's gardens, spot a robin, look at sparkling drops of water on cobwebs. A good, healthy walk is an excellent excuse for another mince pie.

VISITING OTHER TOWNS

Walks in the country are great, but you can make many wildlife discoveries visiting other towns or even new areas of the town in which you live. From north to south or east to west Britain is sufficiently different in climate to have noticeable variations in the natural world; as a general rule the west is wetter, the east drier, the north colder and the south warmer. On a more local level hunt out microclimates: wind tunnels, frost pockets, sheltered walled gardens and bracing sea fronts, all will reveal their secrets if you look carefully.

URBAN WILDLIFE WRITING

When it is cold outside or we are just too tired to don coats and go spotting, we like to settle down with a mug a tea and a book. There is a wealth of sensitive writing that looks at urban nature. One of our favourites is *Darwin Comes to Town* by Menno Schilthuizen, which explores the ways that the fast pace of city life has accelerated evolution in many species of urban nature, from birds who have adjusted the pitch of their song to be heard above traffic, to mice that have adapted metabolisms to cope with the high-fat diet from discarded junk food. If we are leaning towards fiction, then we might dip into Patrick Süskind's *The Pigeon*, in which a security guard is driven towards madness by a pigeon roosting opposite his apartment; or Kenneth Allsop's *Adventure Lit Their Star*, the story of some little ringed plovers nesting in an abandoned gravel pit and one man's attempt to save their nest from a determined egg collector.

YEW

This is Britain's longest-lived native tree, easily reaching 600 old years and often living for over a thousand. It is hard to age exactly, as the wood is rotted by a fungus, making many old trees hollow. The flat, dark green needles have surprisingly sharp points, and the small, green flowers ripen into flat-ended red berries, each of which contains a single seed. Every part of the tree is poisonous, apart from the flesh of the berries, which many birds enjoy while not digesting the harmful seeds. The trees are often found in churchyards, but many predate the churches and though they are regarded as symbols of life, no one quite knows why; possibly their longevity is linked with immortality.

PRINTS IN THE SNOW

Prints in the snow are a perfect excuse for unleashing your Davy Crockett and doing a bit of tracking (hat optional). For city dwellers the snow can also offer an opportunity to confirm the existence of those creatures you think are there but never manage to catch sight of in normal busy life. Take some time to examine the tracks – even if you don't quite know what they are, where do they seem to be going to or coming from? Are they bounding or scampering? Looking for something to eat? There can be tracks across pristine snow in the most surprising places – on car roofs or across normally busy intersections. An abstract pattern of tracks can make the most beautiful photograph.

LOOKING BACK AND LOOKING FORWARD

Today is the day to take stock and plan practically for the future, forgetting worthy resolutions you will break. Your nature diaries will almost certainly not be a perfect record of everything you have seen, but looking back over the year or years will bring back memories. More diaries and you can compare different years: did migrating birds arrive earlier? Was the weather worse? How good were the berries? Many organizations depend on information of this type; you can join the volunteers of a number of societies depending on your particular interests. Now is also the time to plan: bluebell woods to visit, canals to walk along, rivers to clean, wildlife to seek out. That is the joy of nature watching; even in the little park at the end of your road there is always something new to see, some tiny detail to notice which, thanks to the generosity of the natural world, will enhance your day and make everything seem a little brighter.

FURTHER RESOURCES

These are some books, apps and websites we liked and found useful.

GUIDEBOOKS

There are a wide range of very good guidebooks available. Choose which
best suits you in terms of layout, level of detail and size, asking yourself if
you want it to fit in a pocket or use an app on the move and consult a more
detailed book at home.

The National Garden Scheme, *The Garden Visitor's Handbook*, published annually

Chinery, Michael, *British Insects: A Photographic Guide to Every Common Species*,
    HarperCollins, 2009

Hume, Rob et al, *Britain's Birds (2nd edition)*, Princeton University Press, 2020

Johnson, Owen and David More, *British Tree Guide*, HarperCollins, 2015

Newland, David *Britain's Butterflies*, Princeton University Press, 2020

Phillips, Roger *Mushrooms,* Pan Macmillan, 2006

Pretor-Pinney, Gavin, *A Cloud A Day*, Batsford, 2019

Sterry, Paul, *Collins Complete Guide to British Wildlife*, HarperCollins, 2008

Streeter, David et al, *British Common Wildflower Guide*, HarperCollins, 2015

NON-FICTION BOOKS

These are some of the many books about nature, whether biographical,
instructive, or descriptive, that we have enjoyed and found useful. Some
may not be strictly urban but encompass a way of looking at and interacting
with nature that works wherever you are. We have given the original
publication dates to put the books into context. For example. Gerard's
Herball is a great read but not a particularly useful identification guide.

Babbs, Helen, *Adrift: A Secret Life of London's Waterways*, Icon Books, 2016

Barnes, Simon, *Rewild Yourself,* Simon & Schuster, 2018

Beer, Andy, *Every Day Nature*, National Trust Books, 2020

Bennett, Victoria, *All My Wild Mothers*, Two Roads, 2023

Cooke, Jack, *The Tree Climber's Guide*, HarperCollins, 2016

Culpeper, Nicholas, *The Complete Herbal and English Physician*, 1653

Gerard, John, *Herball, or Generall Historie of Plantes*, 1597

Grigson, Geoffrey, *The Englishman's Flora*, Phoenix House, 1960

Kilvert, Francis, *Diaries: 1870-1879*

Lewis-Stempel, John, *Nightwalking*, Doubleay, 2022

Lindo, David, *The Urban Birder*, New Holland, 2011

Mabey, Richard, *The Unofficial Countryside*, Collins, 1973

Macfarlane, Robert, *The Wild Places*, Granta Books, 2007

Ed. McMorland Hunter, Jane, *A Nature Poem for Every Day of the Year*, Batsford, 2018

Ed. McMorland Hunter, Jane, *A Nature Poem for Every Night of the Year*, Batsford, 2020

Ed. McMorland Hunter, Jane, *Nature Writing for Every Day of the Year*, Batsford, 2021

Maiklem, Lara, *Mudlarking; Lost and Found on the River Thames*, Batsford, 2020

Moss, Stephen, *The Accidental Countryside*, Guardian Faber, 2020

Reynolds, Mary, *The Garden Awakening*, Green Books, 2016

Schilthuizen, Menno, *Darwin Comes to Town*, Quercus, 2018

Sheldrake, Merlin, *Entangled Life*, Vintage, 2021

Tree, Isabella, *Wilding*, Picador, 2018

White, Gilbert, *The Natural History and Antiquities of Selborne*, 1789

Wilkinson, Florence, *Wild City*, Orion, 2022

FICTION
There are many great novels in which nature plays a leading role.
Here are a few of our favourites.

Allsop, Kenneth, *Adventure Lit Their Star*, Penguin, 1949

Calvino, Italo, *The Baron in the Trees*, 1957 Italian, 1959 English

Harrison, Melissa, *Clay*, Bloomsbury, 2013

Macauley, Rose, *The World My Wilderness*, Virago, 1950

Powers, Richard, *The Overstory*, Vintage, 2018

Shafak, Elif, *The Island of Missing Trees*, Penguin, 2022

Süskind, Patrick, *The Pigeon*, 1987 German, 1989 English

WEBSITES

ANIMALS, INSECTS AND REPTILES
arguk.org (Amphibian and Reptile
Groups of the UK)
bigbutterflycount.butterfly-
conservation.org
wildlifetrusts.org

BIRDS
rspb.org.uk
wwt.org.uk
(Wildfowl and Wetlands Trust)

CLIMATE AND WEATHER
metoffice.gov.uk
rmg.co.uk (Royal Observatory Night Sky)
theskylive.com

LAND AND WATER
canalrivertrust.org.uk
righttoroam.org.uk

PLANTS
commonground.org.uk/apple-day
forestresearch.gov.uk
(Tree Alert scheme)
guerillagardening.org
kabloom.co.uk
ngs.org.uk (National Garden Scheme)
nsalg.org.uk
(National Allotment Society) plantlife.
org.uk
theorchardproject.org.uk
treecouncil.org.uk

wildaboutgardens.org.uk
woodlandtrust.org.uk

OTHER
earthday.org
nationaltrust.org.uk
nhm.ac.uk (National History Museum)

APPS

TO IDENTIFY AND RECORD YOUR FINDS
BirdTrack
(British Trust for Ornithology)
Big Butterfly Count
Cloud Spotter
Collins Bird Guide
iNaturalist
Picture Insect
Smart Bird ID (RSPB)
Tree ID (Woodland Trust)

FOR DAYS OUT OR ACTIVITIES
My Pollen Forecast
Nature Finder (Wildlife Trust)
NGS Find A Garden
(National Garden Scheme)
National Trust-Days Out

THE AUTHORS

Jane McMorland Hunter is a
writer, bookseller, gardener
and cook. She works at
Hatchards bookshop and has written twelve books for
Frances Lincoln, The National Trust, Teach Yourself, Pavilion and Prospect
and has edited twelve anthologies for Batsford and the National Trust. She
is a trustee of the Scottish Rainforest Education Centre, based in the very
non-urban wilds of Wester Ross in Scotland.

Sally Hughes is a writer, bookseller and cook. She works at
Waterstones and has edited several cookery compendiums for Books
for Cooks.

Jane and Sally discovered a shared love of the natural world when
they were working together in a London bookshop where they spent their
lunch hours picnicking in the local park, sheltering under the bandstand
when necessary. Their boss got used to them returning late from breaks
because they'd 'spotted something amazing'. Together, they have written
three books in 'The English Kitchen' series for Prospect books on Berries,
Nuts and Cherries and Mulberries.

ACKNOWLEDGEMENTS

We would both like to thank everyone at Hatchards (Piccadilly, St Pancras and Cheltenham) and Waterstones for the support they have given to all our books and for being so tolerant and encouraging to the writers in their midst.

We would also like to thank Nicola Newman and Hattie Grylls at Batsford, our wonderful editors.

When we wrote the cookery and gardening books for 'The English Kitchen' series, all our family and friends obligingly ate and admired the various fruits we were writing about. Since then, they have indulged our obsession with nature, allowing it to take over everything from shopping trips to picnics in the park. Some have even been swept up by our enthusiasm and can be found peering at the ground saying, 'I think it's a...'. Thank you to all.

Jane: In particular, I'd like to thank Mat for telling me not to move to the countryside. Also, Matilda, my grey tabby cat, for being a dedicated paperweight and enthusiastic chaser of paper.

Sally: I would like to thank my parents, who dragged me, protesting loudly, away from my books to go on family nature rambles until I learned to love them for myself.

POETRY SOURCES

'The Natural World', George the Poet © George the Poet. Reprinted with kind permission of Brotherstone Creative Management.

'Birmingham Roller', Liz Berry © Black Country by Liz Berry, 2014, Chatto & Windus

'The Trees', Philip Larkin © The Complete Poems by Philip Larkin, 2012, Faber and Faber

'Drought', David Holbrook © Against the Cruel Frost by David Holbrook, 1963, Putnam Publishers

'Love Song, 31st July', Richard Osmond © Useful Verses by Richard Osmond, 2017, Picador

'Creative Waiting', Roger McGourgh © Safety in Numbers by Roger McGough, 2021, Viking

INDEX